Aromatic Chemistry

Malcolm Sainsbury

School of Chemistry, University of Bath

OXFORD NEW YORK TOKYO
OXFORD UNIVERSITY PRESS
1992

Oxford University Press, Walton Street, Oxford OX2 6DP

Oxford   New York   Toronto
Delhi   Bombay   Calcutta   Madras   Karachi
Petaling Jaya   Singapore   Hong Kong   Tokyo
Nairobi   Dar es Salaam   Cape Town
Melbourne   Auckland
and associated companies in
Berlin   Ibadan

Oxford is a trade mark of Oxford University Press

Published in the United States
by Oxford University Press, New York

A catalogue record for this book is available from the British Library

Library of Congress Cataloging in Publication Data
Sainsbury, M.
Aromatic chemistry / Malcolm Sainsbury.
p. cm.— (Oxford chemistry primers ; 4)
1. Aromatic compounds. I Title. II. Series.
QD331.S25 1992   547'. 6—dc20   91–47606

ISBN 0 19 855675 6   (Hbk)
ISBN 0 19 855674 8   (Pbk)

Typeset by Pentacor PLC, High Wycombe
Printed in Great Britain by
Information Press, Oxford

Series Editor's Foreword

The study of the structure and properties of benzene and the development of the concept of aromaticity represents a significant part of the foundation of modern organic chemistry. Very many natural products may be classed as aromatic compounds and the development of the chemical industry has relied heavily on aromatic chemistry. It is not surprising, therefore, that an understanding of aromatic chemistry forms a crucial part of all modern chemistry courses.

Oxford Chemistry Primers have been designed to provide concise introductions relevant to all students of chemistry and contain only the essential material that would usually be covered in an 8–10 lecture course. In this primer, Malcolm Sainsbury provides an excellent account of the fascinating topic of aromatic chemistry presented in an easy to read and student-friendly style. This primer will be of interest to apprentice and master chemist alike.

Stephen G. Davies
The Dyson Perrins Laboratory, University of Oxford

Preface

The chemistry of benzene, its derivatives, and its analogues is a vital part of organic chemistry and a core subject in undergraduate programmes in chemistry, biochemistry, and pharmacy. It is a vast subject area, and the aim of this book is to provide an introduction which deals with the fundamentals and provides the basis for deductive reasoning. In addition, it emphasizes and exemplifies the interrelationship of aromatic to aliphatic chemistry; thus, whenever appropriate, comparisons between the behaviour of aromatic and aliphatic compounds are highlighted and discussed. The concept of aromaticity itself is explained and its wider implications to heterocyclic chemistry and the reactivity of polyunsaturated small and large cyclic ring systems are introduced. All the major reactions of aromatic chemistry are interpreted, and consideration is given to the determination of orientation in aromatic substitution reactions. Modern synthetic methods such as those involving the uses of organometallic reagents are included and the importance of arenes and carbenes in aromatic chemistry is discussed. Concepts such as kinetic versus thermodynamic control of reactions, reagent approach control, and acidity and basicity are all featured. In short, the book summarizes most of aromatic chemistry and its content is based upon the author's detailed lecture notes to first and second year students, compiled during thirty years of teaching and research in this area of science.

I would like to thank my colleague and friend Dr David Brown for many helpful comments and constructive criticisms concerning the contents and presentation of this book.

Bath
November 1991

M.S.

Contents

Oxford Chemistry Primers

1 S. E. Thomas *Organic synthesis: the roles of boron and silicon*

2 D. T. Davies *Aromatic heterocyclic chemistry*

3 P. R. Jenkins *Organometallic reagents in synthesis*

4 M. Sainsbury *Aromatic chemistry*

5 L. M. Harwood *Polar rearrangements*

6 I. E. Markó *Oxidations*

7 J. H. Jones *Amino acid and peptide synthesis*

8 C. J. Moody and G. H. Whitham *Reactive intermediates*

---

1 Aromatic character and the structure of benzene

1.1 Introduction

Benzene C_6H_6 is a component of coal tar distillate and an aerial pollutant found in the exhaust gases of automobiles. It is the parent of a group of compounds known as *arenes* which exhibit different reactions to those shown by other unsaturated hydrocarbons, i.e. *alkenes*. Arenes are also described as aromatic molecules. The term aromatic is historical and derives from the fact that many naturally occurring fragrant compounds were observed to contain a benzene unit. Nowadays aromaticity has a wider scientific meaning and refers to planar cyclically conjugated structures having $(4n+2)$ π electrons (where n is 0, 1, 2 etc.), **Hückel's rule**.

Early chemists accustomed to the concept of the tetravalency of carbon had great difficulty in interpreting the structure and chemical behaviour of benzene. Thus, for example, Ladenburg proposed that benzene had a prismatic structure, whereas Dewar considered it to be a distorted hexagon with a long central bond and two double bonds. A more reasonable interpretation was that of the German scientist Kekulé, who suggested that the molecule had a cyclic arrangement of carbon atoms joined by alternate single and double bonds.

Ladenburg   Dewar   Kekulé

Fig.1.1 Some early structures for benzene

1.2 Resonance and the π electron system of benzene

Although the Kekulé hypothesis was an advance in understanding, it did not explain, for example, why there is only one 1,2-dichlorobenzene, two isomers of which should exist, one with the carbons bearing the chlorine atoms separated by a double bond, and the other with them separated by a single bond. Furthermore, single bonds (154 pm) are longer than double bonds (133 pm) and this would give rise to an irregular structure. This criticism was countered by Kekulé who argued that a very rapid equilibrium existed between, say, the two molecules of 1,2-dichlorobenzene, thereby averaging out the bond lengths (Fig. 1.2).

Fig.1.2 Kekulé structure for dichlorobenzene

One effect of the continuous π system is to create a ring current, which in turn generates a magnetic field. In the ^{1}H NMR spectrum the ring protons of benzene give rise to a single peak at δ 7.2. This chemical shift value is at lower field than expected for the signals of alkenic protons (typically δ 5.8–6.5): this down-field shift is caused because the ring current enhances the applied field in the plane of the ring where the protons reside. Similar ring currents are produced for other planar cyclic polyenes containing $(4n+2)$ p electrons which also show proton signals close to δ 7.0. ^{1}H NMR spectroscopy can thus be used as a simple probe of aromatic character (see p. 87).

This is incorrect: no such equilibration exists and there is only one 1,2-dichlorobenzene. In benzene six carbon atoms each bond to a single hydrogen atom and to two other carbons forming a symmetrical sigma-bonded framework. X-ray studies confirm this and show the ring to be a regular hexagonal structure in which all carbon-bonds are equal (140 pm) with bond angles of 120°. Each trigonal (sp^2-hybridized) carbon atom nominally has a single p orbital containing one electron. Since all the carbon–carbon bond lengths are identical, individual p orbitals must overlap equally with their neighbours so that a continuous π system is created above and below the ring within which the p electrons circulate (Fig 1.3).

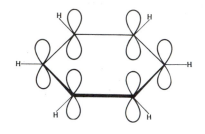
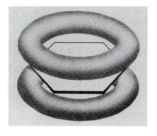

Fig. 1.3 Overlap of p orbitals in benzene

1.3 Representation of benzene

While the circle device is attractive for monocyclic arenes its application to polycyclic molecules is strictly incorrect. A full circle within a ring signifies a π system containing six electrons. Thus in the structure of the fused bicyclic arene naphthalene (see p. 70, for example) such a representation would imply that there were twelve π electrons, six in each ring, whereas in fact there are only ten! Great care has to be exercised when using this device, especially as full circles are also commonly drawn in representations of monocyclic ions which may contain more, or less, than six π electrons, see p. 4. Some authors recommend dashed circles to describe π systems other than those containing six electrons.

How, then, can we represent benzene, which contains a fully delocalized sextet of p electrons? One way is to draw a hexagon with a circle inside it to symbolize the free circulation of the six p electrons. A second representation uses the **valence bond** approach: here as many classical (electron localized) structures as are necessary to describe the molecule are drawn, thus benzene is regarded as a **resonance hybrid** of the two Kekulé formulae. Other planar representations are also included since they reflect the possibility of p electron overlap across the ring (Fig 1.4). It is important to realize that *no one formula truly represents benzene* and, since the positions of the atoms are fixed, *only the electrons are free to move*. In the valence bond approach each contributing form is regarded as a factor in the overall wave equation of the benzene molecule. Such an equation does not exist without all its component factors, but some are more important than others. Thus since overlap across the ring is less likely than the overlap between p electrons on adjacent carbon atoms, the Dewar contributions are less important than the Kekulé types. A double-headed arrow between each contributor, or canonical form, is used to indicate their interdependence in the description of a single molecule, *but care must be exercised not to confuse this symbol with that used to indicate equilibration ($\rightleftharpoons$) between different molecules.*

Fig. 1.4 Valence bond representaions of benzene

Chemists tend to use 'curly' arrows to show electron movements and to summarize reaction mechanisms. For this purpose classical structures are required and it is customary to use one Kekulé formula to represent benzene, its limitations being taken for granted. This is the system adopted in this book.

1.4 Resonance energy and the importance of Hückel's rule

Benzene is more stable than predicted: for example, when cyclohexene is hydrogenated 120 kJ mol^{-1} of energy is released. It follows that for a hypothetical cyclohexatriene c. 3 x 120 = 360 kJ mol^{-1} of heat *might* be generated. When benzene itself is hydrogenated the heat actually liberated is 210 kJ mol^{-1} (Fig. 1.5).

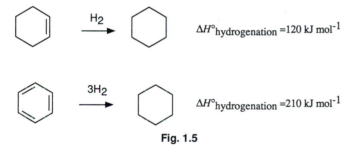

$\Delta H°$hydrogenation =120 kJ mol^{-1}

$\Delta H°$hydrogenation =210 kJ mol^{-1}

Fig. 1.5

The difference between the calculated and the experimentally derived values ($3 \times 120 - 210 = 150$ kJ) is called the **empirical resonance energy**, and it illustrates the extra stability that benzene has over a hypothetical irregular cyclohexatriene molecule. This is, however, not the *true* resonance, or delocalization, energy since this would be the difference in energy between benzene and a symmetrical non-delocalized cyclohexatriene unit, which does not exist and for which there is no simple predictive model. As a result of this uncertainty, theoretical calculations of the resonance energy of benzene vary from 40 to 120 kJ mol^{-1}. Whatever the actual figure, there is no doubt of the stabilizing effect of electron delocalization provided by the formation of the delocalized π system in benzene. Addition reactions, such as hydrogenation, are thus much less easy than might have been expected. In fact the chemistry of benzene is dominated by substitution reactions in which the products retain the π system and only the hydrogen atoms are replaced by other groups.

Importantly, the empirical resonance energy of benzene is greater than that of its *acyclic* analogue hexatriene, and the latter molecule, like others in its class, undergoes addition rather than substitution reactions. Clearly, delocalization of the *p* electrons over six carbon atoms alone does not account for the special properties of benzene. A *cyclic array* is necessary, but if this were the only factor then the higher

Cyclooctatetraene

Cyclooctatetraene dianion

It might have been supposed that the chemical shift of this resonance should be at *c.* δ 7.2, comparable with the proton resonance of benzene; the discrepancy is thought to be due to the effect of the charge which acts to oppose the induction of the ring current.

homologue of benzene cyclooctatetraene would also be expected to show aromatic properties. Cyclooctatetraene is a stable compound, but unlike benzene it is not planar and it exists in the form of a 'tub'. Two types of carbon–carbon bond are present, four in which the atoms are separated by 150 pm, and four by 135 pm.

These bond lengths are closely similar to those of single and double bonds, respectively, in simpler molecules. The molecule undergoes addition reactions rather than substitution, and therefore behaves as a typical alkene. Furthermore, there is no significant difference between the calculated and the experimentally determined energy content of cyclooctatetraene. However, cyclooctatetraene can be converted into a dianion which, significantly, has a planar structure: all the carbon-carbon bond lengths are equal (141 pm), and it has a delocalized π system. In the ¹H NMR spectrum the protons of the dianion are all equivalent and give rise to a singlet at δ 5.7.

Cyclooctatetraene and its dianion differ in the number of *p* electrons present, and, as we shall see, this determines whether 'aromatic character' is observed or not. A related situation is noted for the lower homologue of benzene, cyclobutadiene. Here the parent molecule is unstable and decomposes above 4 K. It might be argued that the instability of this molecule is due to steric problems created by the small ring size and that this outweighs any benefits derived from the delocalization of the π electrons. Ring strain in such a small molecule is undoubtedly important, but the tetramethylcyclobutadienyl dication is stable in liquid sulphur dioxide solution, and can be formed by treating 3,4-dibromotetramethylcyclobutene with antimony pentafluoride in this solvent (eqn 1.1). (¹H NMR studies show that all the methyl groups in the dication are equivalent.)

Cyclobutadiene

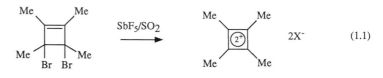

$$SbF_5/SO_2 \qquad\qquad 2X^- \qquad (1.1)$$

Tetramethylcyclobutadienyl dication

Cyclooctatetraene and cyclobutadiene contain eight and four *p* electrons respectively (4*n* systems), whereas benzene has six, and the dianion of cyclooctatetraene has ten (a 4*n*+2 system in which *n* = 2). The dication of cyclobutadiene is an example of a 4*n*+2 species in which *n* = 0.

The presence of (4*n*+2) electrons conveys some aromatic stability to the cyclobutadienyl dication and similarly allows the cyclooctatetraene dianion to achieve planarity. The question then remains: what is special about the (4*n*+2) *p* electron number?

From the results of quantum mechanical calculations Hückel proved that the energy levels of the individual molecular orbitals within the π systems of regular planar-polygonic molecules, with an even number of atoms, form a symmetrical pattern as shown in Fig. 1.6. There is a single lowest energy level with the others arranged above it in pairs of equal energy (degenerate), until a single highest energy level is reached. The number of orbitals will depend upon the ring size of the polygon; some will be bonding and others antibonding. In certain cases a non-

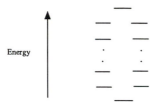

Fig. 1.6 Hückel energy levels for a regular polygonic molecule with an even number of carbon atoms

bonding energy level will also be present: which by definition lies half way between the highest bonding and the lowest antibonding levels.

A representation of the number, the types, and the relative separations of the energy levels for a particular polygon can be obtained by inscribing it inside a circle such that one vertex is at the lowest point of the circle. A line drawn horizontally across the centre of the circle represents the non-bonding level. The energy levels of the molecular orbitals are then indicated by those points where the polygon and the circumference of the circle meet. Those below the non-bonding line are bonding, those above are antibonding. This mnemonic device (see Fig. 1.7), first introduced by Musulin and Frost, not only describes the available molecular orbitals within *potentially* aromatic monocyclic systems, but also illustrates their relative energies.

Antibonding
Non-bonding
Bonding

Fig. 1.7 Frost–Musulin diagrams for 6-, 8-, and 4-membered polygonic molecules

Benzene is a regular planar hexagon so it has three bonding molecular orbitals and three antibonding orbitals. Since there are six *p* electrons to be accommodated, benzene has all its bonding orbitals filled, and it thus achieves stability in the same way as an inert gas. For an eight atom planar polygon there are also three bonding orbitals, but additionally two degenerate non-bonding orbitals. Were this to represent the energy levels in cyclooctatetraene, six of the eight available *p* electrons would occupy the bonding orbitals with their spins paired; but following the stipulation that electrons enter orbitals of equal energy singly, before pairing takes place (Hund's rule), the remaining two electrons must occupy the non-bonding orbitals, *one in each*. These two electrons have their spin states parallel. Planar cyclooctatetraene is thus represented as a triplet diradical. Small wonder then that it is not a planar molecule, since if it were, and a delocalized π system were forced upon it, the two electrons would be forced into close proximity and would strongly repel one another. The analysis of a four membered polygon gives a similar result: if four *p* electrons were present two would occupy the bonding molecular orbital and the other two would be located in non-bonding orbitals with their spins parallel. It follows that fully delocalized cyclobutadiene, like planar cyclooctatetraene, should be a triplet diradical (see Fig. 1.9).

Although orbitals may be degenerate, i.e. have the same energy, they must differ in symmetry properties. This ensures that thay are distinct from one another, in a quantum mechanical sense.

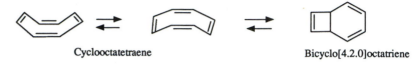

Cyclooctatetraene   Bicyclo[4.2.0]octatriene

Fig. 1.8

Cyclobutadiene

Note that although calculations
suggest that derivatives of the
cyclobutadienyl dication (such as
the tetramethyl salt discussed
above) cannot be completely planar
because of steric interference
between the substituents, there is
sufficient effective overlap between
the *p* orbitals to provide a significant
stabilizing effect.

The cyclooctatetraene molecule is non-planar. it exists as two interconverting forms requiring inversion of the ring, and also exhibits valence isomerism where the single and double bonds can interchange; in addition there is the possiblity of isomerism between the monocyclic structure and bicyclo[4.2.0]octatriene (Fig. 1.8).

The small ring size of cyclobutadiene does not permit it to assume a nonplanar structure. However, there is evidence that below 4 K it exists as a rectangular structure, with alternate single and double bonds, which also rapidly equilibrates between two possible valence isomers. The non-symmetrical arrangement is adopted to overcome the mutual overlap of the two spin parallel p electrons predicted for a molecule having square planar geometry. By adopting a rectangular shape the degeneracy of the highest occupied orbitals is destroyed. Two new orbitals are created, one of lower energy than the other. Now *both* electrons can be accommodated in the lower energy orbital with their spins paired.

Planar cyclic polyenes with $(4n+2)$ p electrons ($n = 0, 1, 2, 3$, etc) fulfil the Hückel requirement for aromaticity and are more stable than their acyclic counterparts (see p. 3). However, similar molecules with delocalized $(4n)$ p electron systems are inherently unstable and are often said to be *antiaromatic*. When two electrons are added to either cyclooctatetraene or cyclobutadiene, they enter the non-bonding levels with the result that now these, as well as the bonding MOs, are filled (see Fig. 1.9). The anions now contain $(4n+2)$ p electrons and, since all the electrons in such species are spin paired, planarity is possible and aromatic stabilization is achieved.

Antibonding

Non–bonding

Bonding

Fig. 1.9 Electronic configurations of regular planar cyclooctatetraene, cyclobutadiene, and their respective dianions

The cyclopropenyl anion, the cyclopentadienyl cations, and the cycloheptatrienyl anion are all antiaromatic, unstable $4n$ systems, but the cyclopropenyl cation, the cyclopentadienyl anion, and the cycloheptatrienyl cation (better known as the tropylium cation) are $4n+2$ species well represented by many stable salts (Fig. 1.10).

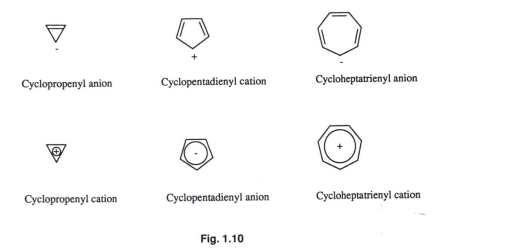

Cyclopropenyl anion	Cyclopentadienyl cation	Cycloheptatrienyl anion
Cyclopropenyl cation	Cyclopentadienyl anion	Cycloheptatrienyl cation

Fig. 1.10

1.5 Nomenclature and numbering

The numbering of monosubstituted benzenes, such as toluene, give the carbon atom bonded to the substituent group (methyl in this case) the lowest number. The numbering system is most commonly used when multiple substituents are present. For monosubstituted benzenes an older system is also widely employed in which the position immediately adjacent to the substituent is denoted *ortho*, the next *meta*, and that furthest away *para*. The carbon atom directly bonded to the substituent is designated *ipso*.

Monocyclic arenes have the general formula C_6X_6, where X can be a variety of atoms or groups, not necessarily the same. The general name for the unit C_6X_5- is aryl, but in the special case of benzene C_6H_6 where X = H, the unit C_6H_5- is called the phenyl group and carries the abreviation Ph (or less commonly Φ). Rather confusingly, $C_6H_5CH_2-$ is known as the benzyl group; its parent is methylbenzene, and it carries the symbol Bn.

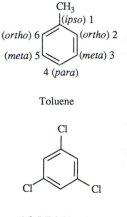

Toluene

1,3,5-Trichlorobenzene

o-Chlorotoluene

2 Reactions of arenes

2.1 Electrophilic substitution

If benzene is reacted with a reagent R (eqn 2.1) it is reasonable to conclude that should R be positively charged (an electrophile) the process will be easier than when R is negative (a nucleophile). For example, as the reagent approaches the nucleus it will first come into contact with the electrons of the π system. If the reagent is negatively charged it will be repelled, whereas if it is positively charged it will be attracted. Also, in the substitution reaction, the reagent replaces a hydrogen from the nucleus. Thus if the attacking species is R⁻ then the hydrogen would leave as a hydride ion, whereas if R⁺ is used the hydrogen is replaced as a proton. Hydride ions have a high energy content, whereas the solvated proton is a low energy species. For these reasons electrophilic substitution is commonplace, whereas nucleophilic substitution only occurs under special circumstances.

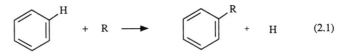

$$+ \quad R \quad \longrightarrow \qquad\qquad + \quad H \qquad (2.1)$$

2.2 Reaction profiles for electrophilic substitution: an addition/elimination process

The energy profile of an electrophilic substitution reaction with benzene as the substrate can be represented diagrammatically, see Fig. 2.1. Here the various steps of the process are illustrated along the horizontal *reaction coordinate*, whilst the vertical axis denotes their relative free energies. Initially, the energy rises as the electrophile (R^+) interacts and disrupts the π system passing through a transition state T_1 in which a sigma bond to one of the ring carbon atoms is partly formed. The difference between the maximum point on the curve and the free energy of benzene represents the activational barrier to the forward reaction $\Delta G^{\#}$. From T_1 the reaction sequence proceeds to form a sigma bonded intermediate (*sigma complex, arenium ion*, or *Wheland intermediate*) (see Fig. 2.2). Here a bond between the electrophile and the ring carbon is fully formed and the hybridization state of the carbon atom at the point of attachment to the electrophile is sp^3, rather than sp^2. The true nature of the transition state is not easily predicted, but it is reasonable to assume that the transition state resembles the sigma complex. However, the energy content of the latter is lowered as some stability is achieved by delocalization of the positive charge over the remaining five sp^2-hybridized carbon atoms of the ring. The close relationship between the transition state and the intermediate is a useful concept, and for

All three names are commonly used and are interchangeable.

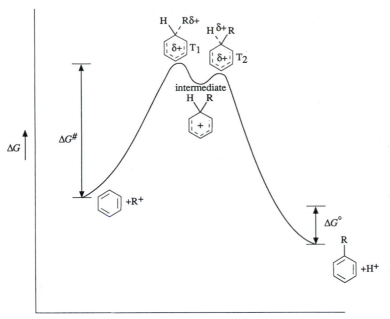

Transition states are transient in nature and cannot be observed directly. As a result it is difficult to gain detailed information about them. In order to overcome the problem it is usual to adopt the **Hammond postulate**: 'if two states, as for example a transition state and an unstable intermediate, occur consecutively during a reaction process and have nearly the same energy content, their interconversion will involve only a small reorganization of molecular structure' (Hammond 1955). Referring to Fig. 2.1, this means that the transition state T_1 is more likely to resemble the intermediate than the reactants. Similarly, the same intermediate is a better model for the second transition state T_2 than the products.

Fig. 2.1 Reaction profile for the reaction of benzene with an electrophile R^+

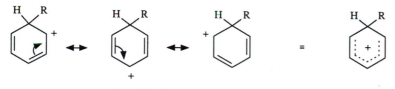

Fig. 2.2 Resonance in a sigma complex

more complex reactions as, for example, the electrophilic substitution of an arene already bearing a functional group, a consideration of the resonance effects within the possible sigma complexes provides an easy means of predicting the preferred orientation of substituents in the product(s) (see p. 18).

In order that the reaction should progress to the final product another transition state T_2 is traversed in which the sigma bond to the proton of the tetrahedral carbon atom weakens and eventually breaks. As this happens there is an initial increase in energy, but this falls as the proton is removed with a base B, and the π system is restored to yield an aromatic product (eqn 2.2).

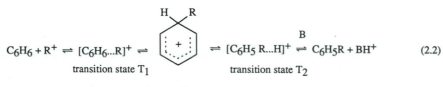

$$C_6H_6 + R^+ \rightleftharpoons [C_6H_6...R]^+ \rightleftharpoons \underset{\substack{\text{intermediate} \\ \text{(sigma complex)}}}{} \overset{B}{\rightleftharpoons} [C_6H_5 R...H]^+ \rightleftharpoons C_6H_5R + BH^+ \qquad (2.2)$$

transition state T_1 \qquad\qquad transition state T_2

In certain cases the whole sequence is reversible and the rates of the forward and back reactions are governed by the heights of the respective activational barriers, represented by the energies of the two transition states T_1 and T_2 respectively. Whether the reaction is energy consuming (endothermic), or energy releasing (exothermic) will depend upon the relative stabilities of the starting compounds and the products. This is reflected in the magnitude of the standard free energy change $\Delta G°$. (If it is assumed that the entropy change $\Delta S°$ for the reaction is small then $\Delta G° = \Delta H° - T\Delta S°$ approximates to $\Delta G° \approx \Delta H°$.)

2.3 Evidence for the intermediacy of sigma complexes

There is much evidence for the formation of sigma complexes. Thus when benzene is treated with SbF_5–HSO_3F in SO_2ClF–SO_2F_2 at -120 °C the benzenenium cation is formed and its ^1H NMR spectrum can be recorded (eqn 2.3). Sometimes sigma complexes can be isolated as solids, as in the case of 1,3,5-trimethylbenzene, which reacts with HF/BF_3 to give 1,3,5-trimethylbenzenenium tetrafluoroborate (eqn 2.4).

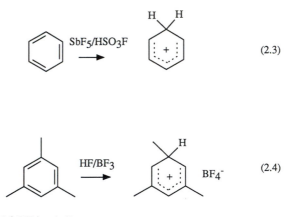

(2.3)

(2.4)

1,3,5-Trimethylbenzene

The fact that sigma complexes form does not prove that they participate in electrophilic substitution reactions, but strong evidence is available from kinetic data. Thus the rates of the nitration of benzene and perdeuteriobenzene, under identical conditions, are virtually the same. If the breaking of the respective C–H and C–D bonds were involved in the rate determining step then differences in the strengths of these bonds would be reflected in different reaction rates. No significant isotope effect is detected and so the rate determining step must be the generation of very similar intermediates, namely two benzenenium ions (eqn 2.5a, b).

In certain types of electrophilic substitution reactions, such as the Friedel–Crafts acylation (see p. 14), small kinetic effects are observed. This does not invalidate the addition/elimination mechanism, rather it indicates that the energy barrier for fragmentation of the carbon–hydrogen isotope bond is greater than that required for the formation of the sigma complex, i.e. $T_2 > T_1$ (see Fig. 2.3).

Calculations suggest a maximum rate difference, k_H/k_D, of about 7 at 25 °C.

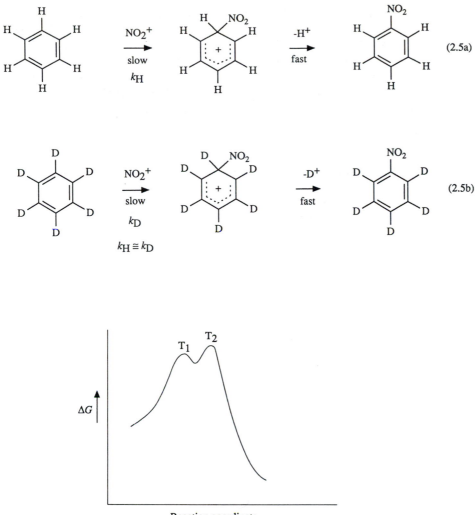

(2.5a)

(2.5b)

$$k_H \cong k_D$$

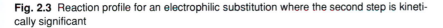

Reaction coordinate

Fig. 2.3 Reaction profile for an electrophilic substitution where the second step is kinetically significant

2.4 π complexes

It is known that π or charge transfer complexes are formed between arenes and a large number of electrophiles (including halogens, Lewis acids, and unsaturated compounds bearing strongly electron withdrawing groups, such as tetracyanoethene and 1,3,5-trinitrobenzene). Thus, for example, the red-brown colour which develops when solutions of iodine (violet) and benzene (colourless) in carbon tetrachloride are mixed is probably due to a π complex in which charge is transferred between the two participants (eqn 2.6).

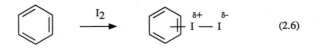

$$(2.6)$$

The recognition that π complexes exist has led some organic chemists to suggest that they may be involved in electrophilic substitution reactions. However, the π bond in such complexes is very weak and their formation and dissociation is so rapid, that even if they do participate it is unlikely that the reaction rates would be affected.

3 Common electrophilic reactions

Reactions such as nitration, sulphonation, halogenation, acylation, and alkylation all proceed through the electrophilic mode and it is useful to understand how the electrophilic reagent is produced in each case.

3.1 Nitration

In nitration reactions the nitronium ion ($^+NO_2$) is the electrophile; this is generated through initial protonation of nitric acid. Any acid stronger than nitric acid will promote the reaction, but concentrated sulphuric acid is the usual choice (eqn 3.1).

Nitronium salts such as the tetrafluoroborate and the perchlorate can also be used as nitrating agents.

$$HNO_3 + H_2SO_4 \rightleftharpoons H_2{}^+NO_3 + HSO_4{}^- \rightleftharpoons H_2O + {}^+NO_2 \qquad (3.1)$$

The nitronium cation attacks the π system forming a sigma complex which is then deprotonated by the counter anion to form nitrobenzene (eqn 3.2). Note that the overall process is catalysed by sulphuric acid.

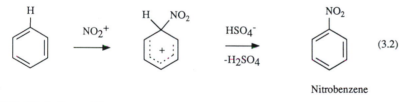

$$(3.2)$$

Nitrobenzene

3.2 Sulphonation

Benzene is sulphonated by heating it with concentrated sulphuric acid at 150 °C. The electrophile is sulphur trioxide (Fig. 3.1). For less reactive arenes oleum (a solution of SO_3 in H_2SO_4) is used.

$$2\,H_2SO_4 \rightleftharpoons SO_3 + H_3O^+ + HSO_4{}^-$$

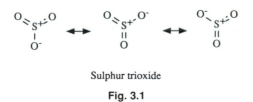

Sulphur trioxide

Fig. 3.1

Although the sulphur trioxide molecule is neutral, the electron withdrawing effect of the three oxygen atoms creates an electropositive centre at the sulphur atom sufficient to allow the reaction to occur readily. Note that the hydrogen sulphate anion which deprotonates the sigma complex thereby forming benzenesulphonic acid arises from the solvent sulphuric acid (eqn 3.3). As a general rule sulphonation reactions are reversible.

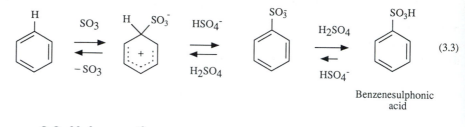

$$(3.3)$$

Benzenesulphonic acid

3.3 Halogenation

Chlorine and bromine enter into substitution reactions with benzene, provided a Lewis acid is present. Typically iron(III) chloride or aluminium(III) chloride is used as the Lewis acid (or carrier) in chlorination, and iron(III) bromide in bromination reactions. The role of the carrier is to form a complex with the halogen, thereby polarizing the homonuclear bond (eqn 3.4a). The 'positive end' of the complex then reacts with the π system affording the iron tetrahalide anion. Deprotonation is effected by this counter anion liberating the halobenzene and restoring the Lewis acid (eqn 3.4b).

$$X\text{-}X \; + \; MX_3 \; \rightleftharpoons \; [X^+...MX_4^-] \qquad (3.4a)$$

halogen Lewis   complex
         acid

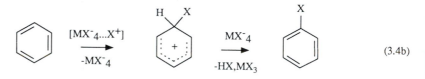

$$(3.4b)$$

X= Cl or Br; MX_3 = $AlCl_3$, $FeBr_3$ etc.

3.4 Alkylation and acylation; Friedel–Crafts and related reactions

Lewis acids are also necessary in alkylation and acylation reactions of benzene with alkyl halides and acyl halides respectively. These are known as Friedel–Crafts reactions. As in halogenation, the carrier polarizes the carbon–halogen bond and is regenerated at the end of the reaction (eqn 3.5a, b). Acid anhydrides can replace acid chlorides in acylation reactions, but here the reaction is expensive because one half of the reagent is 'lost' as the corresponding carboxylic acid (HX = HO_2CR) (Fig. 3.2). A particularly valuable application, however, is the synthesis of aryl-

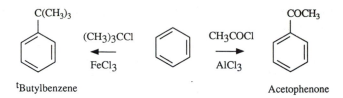

^tButylbenzene   Acetophenone

Fig. 3.2 Examples of Friedel–Crafts reactions

oxoalkanoic acids in which cyclic anhydrides, such as succinic anhydride, are employed (eqn 3.6). In such reactions the whole of the acylating agent is bonded to the arene, and none of the reagent is wasted (Haworth reaction)(see p. 72).

$$R\text{-}X + FeX_3 \longrightarrow \overset{\delta+}{R}\text{-}\overset{\delta-}{X}\text{-}FeX_3 \qquad (R = \text{alkyl or acyl}; X = \text{Cl or Br}) \qquad (3.5a)$$

$$(3.5b)$$

$$(3.6)$$

Alkylation reactions are also possible with alkenes, alcohols, and thioalcohols in the presence of a strong acid. Here the first step is protonation of the double bond, the oxygen atom, or the sulphur atom respectively. The alkylation of benzene with propene is shown in eqn 3.7: the electrophile is the secondary carbocation $(CH_3)_2CH^+$. The product in this case is isopropylbenzene (cumene), see p. 62.

Such reactions can be complicated if rearrangement to a more stable carbocation is possible. This may then become the attacking electrophile and an 'unexpected' alkylarene or a mixture may result (eqn 3.8).

Primary aliphatic carbocations are generally less stable than secondary carbocations, which in turn are less stable than tertiary carbocations.

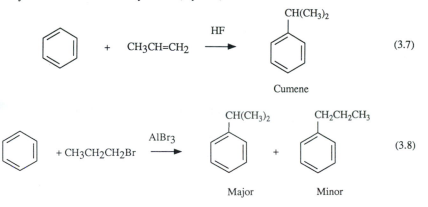

Rearrangements are also possible when alkyl halides are employed since there is considerable carbocation character within the Lewis acid complex (eqn 3.9).

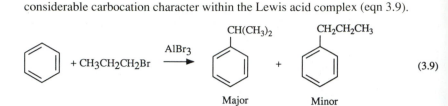

$$CH_3CH_2CH_2\overset{\delta+}{\underset{}{}}\overset{\delta-}{\underset{}{}}...AlBr_4$$

$$\downarrow$$

$$(CH_3)\overset{\delta+}{C}H\overset{\delta-}{...}AlBr_4$$

(3.9)

Formyl chloride and formyl anhydride are unstable molecules, but formyl fluoride in combination with boron trifluoride can be used to formylate arenes. Benzaldehyde is synthesized industrially by heating benzene with hydrogen chloride, carbon monoxide, and aluminium chloride (Gatterman–Koch reaction). Both these reactions are shown in Fig. 3.3.

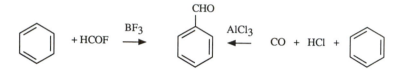

Fig 3.3 Formylation of benzene

Another method of formylating arenes is through the Vilsmeier reaction. This uses *N*, *N*-dimethylformamide and phosphorus oxychloride as reagents. Although the conditions employed are insufficiently forceful to yield benzaldehyde from benzene, the reaction is suitable for the formylation of arenes bearing electron donating groups (eqn 3.10a, b). The weak electrophile in the Vilsmeier is an iminium salt which is generated as shown in eqn 3.10a.

$$(CH_3)_2NCHO + POCl_3 \;\;\overset{-Cl^-}{\rightarrow}\;\; (CH_3)_2N^+{=}CHO(POCl_2) \rightarrow \underset{-PO_2Cl_2^-}{\overset{Cl^-}{(CH_3)N^+{=}CHCl}} \qquad{}^-(3.10a)$$

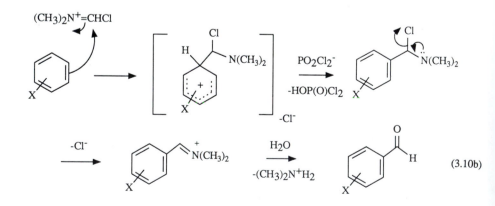

(3.10b)

imine

iminium cation

A related reaction is the Hoesch (Houben–Hoesch) synthesis of aromatic ketones. In this case the reagent is obtained by reacting an arene with a nitrile (RCN) in diethyl ether saturated with hydrogen chloride containing zinc chloride. The initial product is an imine which under the reaction conditions is protonated to give the corresponding iminium salt. This may separate from the reaction mixture and can be hydrolysed in a follow up step. The reaction is most suitable for highly activated arenes, i.e. where X is an electron donating group (eqn 3.11).

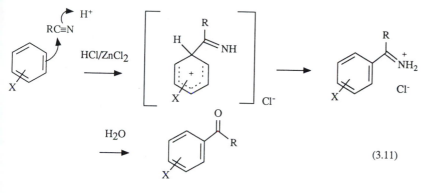

(3.11)

3.5 Electrophilic metallation reactions

Some metals such as mercury and thallium which form covalent carbon–metal bonds react with arenes in an electrophilic sense. Mercuration has not found much use, but thallation with thallium trifluoroacetate is useful in synthesis, because the thallium can be replaced by other groups such as I or CN in a later step (see eqn 3.12). [Iodo- and cyano-arenes are not normally available except via diazonium salts (see p. 57).] The electrophile has not been determined, but could be $Tl^+(O_2CCF_3)_2$. When substituted benzenes are thallated the reaction sometimes shows a high degree of regioselectivity.

The reaction may involve a Meisenheimer type complex, see p. 28, made possible by the presence of the strongly electron withdrawing thallium bearing substituent group.

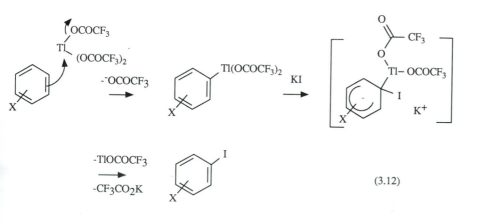

(3.12)

4 Orientation of electrophilic substitution reactions

4.1 The isolated molecule: activating and deactivating groups

Attack may also occur at the *ipso* position, i.e. at the carbon atom to which the group X is already bonded. For electronic or steric reasons this may be unfavourable. In addition, the original substituent X may not form a good cationic leaving group, so that even if the appropriate intermediate is formed it simply collapses back to the starting materials. Despite these factors *ipso* substitution can be a significant reaction under certain circumstances.

When an arene C_6H_5X is substituted by an electrophile R^+ there are three principal sites for bonding: C-2 (*ortho, o*), C-3 (*meta, m*), and C-4 *(para, p)*. From a statistical point of view since there are two *ortho*, two *meta*, and one *para* position the ratio of products should be 2:2:1. However, the nature of the group X has a major influence upon both the site of the attack and the ratio of products formed. Furthermore, it has a decisive effect upon the conditions needed to bring about the reaction. These facts are exemplified by the results of nitration experiments upon some simple monosubstituted benzenes (see Table 4.1).

Table 4.1 Orientation preferences and reaction conditions for the nitration of some benzene derivatives

Substrate C_6H_5X	Product composition			Reaction conditions
	ortho	*meta*	*para*	
X = NO_2	7	88	1	$HNO_3/H_2SO_4/100\ ^\circ C$
X = CH_3	62	5	33	$HNO_3/H_2SO_4/25\ ^\circ C$
X = OCH_3	71	1	28	$HNO_3/Ac_2O/10\ ^\circ C$
X = OH	55	1	45	$HNO_3/H_2O/20\ ^\circ C$

The results in Table 4.1 indicate that it is more difficult to nitrate nitrobenzene than it is to nitrate the other compounds. In addition, whereas the nitration of nitrobenzene gives mainly *m*-dinitrobenzene, the other three compounds give products which arise predominantly through attack at the *ortho* and *para* positions.

4.2 Inductive effects within the isolated molecule

Halogen atoms and most of the commonly encountered substituent groups, apart from alkyl units (which are built up from sp^3 carbon atoms), are more electronegative than the sp^2 hybridized carbon atoms which constitute the benzene ring. As a result a dipole is created and charge is withdrawn from the ring. Inductive effects are exerted through bonds and diminish rapidly with distance: they will thus be strongest at the *ortho* positions and weakest at the *para*. While inductive effects certainly do have an influence on the rates of reactions, their ability to determine the orientation of the incoming electrophile is normally overridden by resonance (delocalization) effects.

4.3 Resonance effects within sigma intermediates

For the present we will assume that the intermediate which forms fastest, i.e. that which has the lowest internal energy, will lead to the predominant product (kinetic control). Table 4.2 shows the relative rates of nitration for some prominent arenes and it is clear that the substituent group, as well as determining the product ratios, also dictates the overall reaction rate. This is mainly a result of resonance within the sigma intermediate and the effectiveness of the substituent already present in assisting, or inhibiting, the delocalization of the positive charge.

Strictly speaking the transition states should be considered, but unfortunately the geometries and charge distributions of such species are not easily assessed. However, as pointed out on p. 9 it is reasonable to assume that a transition state will resemble its related sigma complex.

Table 4.2 Relative rates of nitration of some arenes

C_6H_5X	Relative rate
X = OH	1000
X = CH₃	25
X = H	1
X = I	0.2
X = Cl	0.03
X = NO₂	6×10^{-8}
X = N⁺(CH₃)₃	1×10^{-8}

[N.B. for certain of these processes more than one isomer may form as the product; since the individual rates leading to the different isomers may not be the same, the data refer to the *overall* rate of reaction.]

For example, if the *ortho, meta,* and *para* sigma complexes for the substitution of nitrobenzene with an electrophile R^+ are considered, the internal resonance of the resident nitro group places a positive charge next to the ring. In the *ortho* and *para* complexes this is conjugated with the positive charge carried by the ring carbon atoms. This is not so for the *meta* intermediate where the two charges are separated. The internal energies of the *ortho* and *para* complexes are thus higher than that of the *meta* intermediate and they are formed more slowly (Fig. 4.1). The electrophilic substitution of arenes already bearing substituents containing carbonyl (C=O), sulphonyl (O=SO₂H), cyano (C≡N), and similar electron-withdrawing unsaturated groups, bonded to the ring can be treated similarly and give the same result.

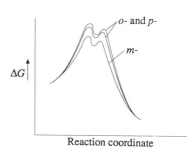

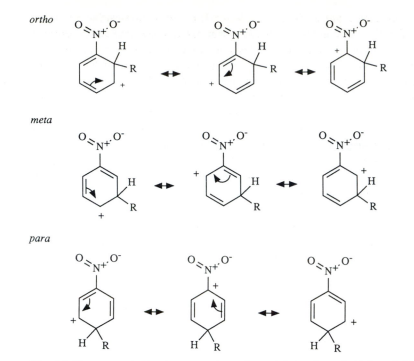

Fig. 4.1 Resonance in the sigma complexes formed during the substitution of nitrobenzene with an electrophile R⁺

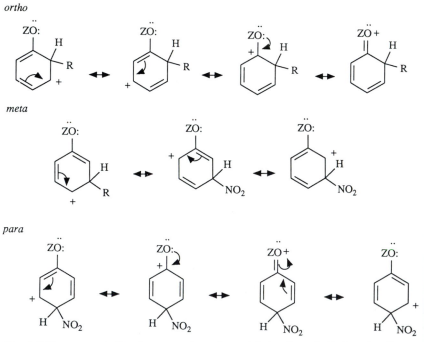

Fig. 4.2 Resonance in the sigma complexes formed during the substitution of methoxybenzene or phenol with an electrophile R⁺

Contrasting results occur for compounds such as methoxybenzene (Z = CH$_3$), or phenol (Z = H)(Fig. 4.2), which have an oxygen atom directly bonded to the benzene nucleus. Here the oxygen atom has a lone pair of electrons in an orbital which is co-planar with and overlaps the π system. In the cases of the *ortho* and *para* intermediates this allows the lone pair electrons to delocalize with the nucleus and to share the positive charge with the oxygen atom. As a result the energy contents of the *ortho* and *para* intermediates are reduced relative to that of the *meta* in which this is not possible. The rates of formation of the *ortho* and *para* intermediates are thus faster than that of the *meta*.

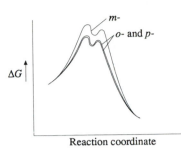

Similar intermediates can be drawn for thiols and the halobenzenes, but note that an increase in atomic diameter decreases the effectiveness of the overlap between the lone pair electrons and the pi system. A lowering of the resonance effect may thus be anticipated in the order O>S and F>Cl>Br>I.

While resonance normally has the dominant role in determining the site adopted by the entering electrophile (**the electromeric effect**), the rate of the reaction is also influenced by the electron withdrawing power of the original substituent (**the inductive effect**). Thus *all* substituents which are more electronegative than hydrogen exact a negative penalty on the reaction rate and the observed kinetics are in fact a compromise between the electromeric and the inductive effects. In chlorobenzene, for example, the lone pairs of electrons on the chlorine atom assist in delocalizing the positive charge in the *ortho* and *para* sigma complexes formed by attack of an electrophile R$^+$, whereas this is not possible for the *meta* complex. This ensures that the preferred reaction products are the *ortho* and *para* derivatives, but since the chlorine atom is strongly electron withdrawing the overall reaction rate is much slower than the reaction of the electrophile with benzene itself (Fig. 4.3).

As an illustration the rate of nitration of benzene is about 30 times faster than that of chlorobenzene, see Table 4.2.

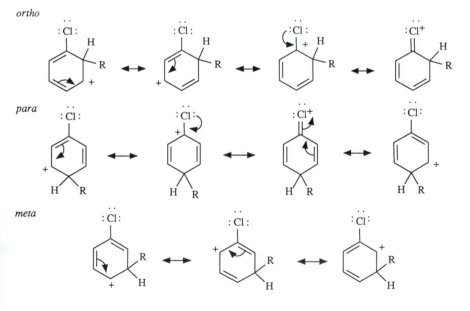

Fig. 4.3 Sigma complexes for the electrophilic substitution of chlorobenzene

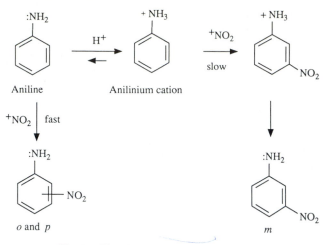

Fig. 4.4 The nitration of aniline in acidic media

The conditions used and the ability of the substrate to act as an acid or a base, also have a fundamental influence on the course of aromatic electrophilic substitution reactions. Thus the amino group of aniline strongly activates the molecule towards electrophilic attack, but aniline is a weak base and protonation, or reaction with a Lewis acid, generates the anilinium cation. Now the lone pair electrons are no longer available to delocalize the charge in the *ortho* and *para* sigma complexes and because the group $^+NH_3$ itself bears a positive charge it is *meta*-directing and deactivating. The rate of *meta* substitution is therefore slow (note the effect of the $^+NMe_3$ group upon nitration quoted in Table 4.2) and in practice the electrophilic monosubstitution of aniline in acidic media leads to a mixture of all three isomers. Thus, although protonation may be almost complete, an equilibrium still exists between the protonated and unprotonated forms, and sufficient aniline reforms to initiate a fast reaction giving *ortho* and *para* substituted products (Fig. 4.4)(see p. 55).

Phenols are weak acids and in alkaline solutions form phenate anions (see p. 63). The fully developed negative charge of the anion may be utilized to stabilize *ortho* and *para* sigma complexes — an advance on the stabilizing effect provided by the lone pair electrons of the parent phenol. As a consequence, reactions with electrophiles are more rapid in alkaline solution than they are in acidic or neutral media (eqn 4.1).

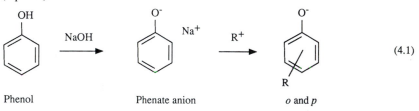

Phenol   Phenate anion   *o* and *p*

(4.1)

As we have seen, alkylbenzenes like toluene (methylbenzene) are activated towards electrophilic attack and *ortho* and *para* substituted products predominate. Here the arenium ions for the *ortho* and *para* reactions possess stabilizing tertiary carbocation character, not available to the *meta* intermediate (Fig. 4.5).

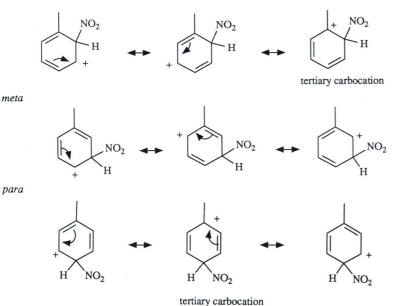

ortho

meta

para

tertiary carbocation

Fig. 4.5 Arenium ions (sigma complexes) from electrophilic attack upon toluene

Ortho/para directing groups in electrophilic substitution reactions include: NR_2, NH_2, OH, O^-, NHCOR, OR, SR, and the halogens (SH is excluded from the list because electrophiles tend to react at the sulphur atom and displace H^+). Alkyl and aryl groups are also *ortho/para* directors, the latter because the π electrons can be used in the same way as a lone pair of electrons to delocalize the charge in the arenium ions. Groups containing electronegative atoms plus an unsaturated bond adjacent to the ring are normally *meta* directors; these include: NO_2, SO_3H, CHO, CO_2H, CO_2R, $CONR_2$, and CN. Positively charged substituents, e.g. $^+NR_3$, $^+SR_2$, $^+PR_3$, or strongly electron withdrawing groups, e.g. CCl_3 or CF_3, are also *meta* directors.

4.4 *Ipso* substitution

The protonation of 1,3,5-trimethylbenzene by HF/BF_3 (see p. 10) is an example of a reaction in which the attacking electrophile bonds to a carbon atom already bearing a substituent group. This is called *ipso* attack. In this case the cation produced is relatively stable, but in others, where the original substituent may support a positive charge or where it can be deprotonated, overall substitution occurs (e.g. eqn 4.2).

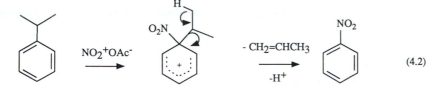

(4.2)

Alternatively an intramolecular rearrangement may occur and a group at the *ipso* position may migrate to another site. Here the result is similar to that obtained by 'normal' electrophilic attack at an unsubstituted carbon atom, but the product ratios may differ. The arenium ions in *ipso* reactions can sometimes be trapped by available nucleophiles. Thus in the case of the nitration of 1,2-dimethylbenzene with nitric acid–acetic anhydride the dimethylbenzenenium ion reacts with acetate anion to give a dihydrocyclohexadiene adduct. 1,4-Elimination of nitrous acid then affords 4-acetoxy-1,2-dimethylbenzene (eqn 4.3).

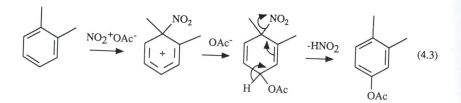

$$(4.3)$$

4.5 Kinetic versus thermodynamic control

For a reaction capable of forming two or more products, the above discussion would indicate that the product ratio would follow directly from the relative stabilities of the individual sigma complexes, or, more correctly, from the energies of the respective activation barriers, i.e. the more accessible the intermediate the more abundant the corresponding product. However, this is commonly not observed and, so far, we have not considered the thermodynamic stabilities of the final products. Thus although one compound may form faster than another, it may involve less strong bonds, or it may be more sterically crowded. An important point to remember is that all reactions are potentially reversible, provided enough time and energy is available, and product ratios are often radically changed simply by raising the temperature and/or extending reaction times.

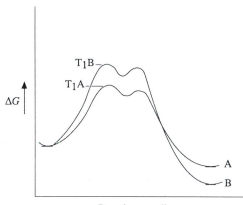

Fig. 4.6 Profile of a reaction in which product A is kinetically favoured, but product B is thermodynamically more stable

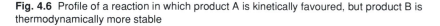

The situation is illustrated in Fig. 4.6 where two products A and B are formed from the same starting material. Two reaction profiles are shown. One leads to product A which has a lower activation barrier than product B and is thus kinetically preferred. Product B, on the other hand, is more stable, and if the reaction was carried out at increasing temperatures or over a longer period the proportion of product B would increase at the expense of A as the latter is recycled and the system approaches equilibrium.

In the sequel a number of examples of kinetic versus thermodynamic control will be found. A classic of its kind is the sulphonation of naphthalene (p. 74).

4.6 Birch reduction

Although the catalytic hydrogenation of benzene and its derivatives is difficult because of the loss of resonance stabilization (see p. 3), it is possible to effect the reduction of many arenes to 1,4-dihydroarenes, including benzene itself, by the Birch reduction. In this procedure the arene, in liquid ammonia containing an alcohol (XOH) as a source of protons (ammonia is insufficiently acidic), is treated with a metal (M), usually lithium or sodium. Initially, a single electron transfer occurs from the metal to the arene to give a radical anion which is then protonated. The radical formed receives another electron yielding an anion, which protonates to form the 1,4-dihydroarene. If an aliphatic amine is the solvent, rather than ammonia, the reaction may continue so that two, or all three, 'double bonds' of the parent arene are reduced.

In the case of alkylbenzenes the rate of reaction is slower than for benzene itself. This is due to electron donation from the substituent, through the hyperconjugative effect. The retarding effect is further enhanced for other groups, e.g. OAlkyl or N(Alkyl)$_2$, in which a pair of electrons can be conjugated with the π system of the ring. For these substrates the products are 2-substituted 1,4-dihydrobenzenes (1-substituted 1,4-cyclohexadienes) (eqn 4.4).

One (*not totally convincing*) explanation for the regioselectivity observed in the Birch reduction of arenes with an electron donating substituent is that the negative charge of the radical anion 'avoids' the electron supply of the substituent. The

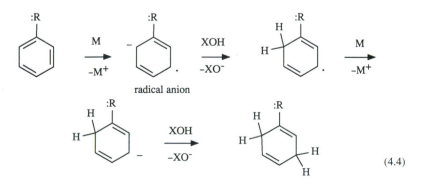

(4.4)

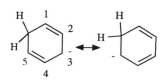

converse would then be true for those arenes bearing electron withdrawing groups (e.g. CO_2H, or $CONR_2$). Certainly the reaction rate of the first step is now faster, and the electron withdrawing substituent is found on a sp^3-hybridized carbon atom in the product. Regardless of the nature of the substituent, the hydrogen atoms are normally added *para* to each other, i.e. 1,4-cyclohexadienes are formed in preference to 1,3-cyclohexadienes, although the latter are more conjugated. This is a puzzle! Indeed, why in, say, the Birch reduction of benzene does the final anion, a resonance hybrid, protonate at C-3 rather than at C-5 (or C-1)? This can be explained by invoking Hine's principle of least motion (Hine 1977): 'reactions which involve the least change in atomic position and electronic configuration are normally favoured'. In this case the argument is rather complex, and rests on the lower number of valence bond order changes required for 1,4-addition (see March 1985).

5 Nucleophilic aromatic substitution

5.1 Second-order reactions

In the case of alkyl halides two main modes of nucleophilic substitution are possible. For those halides which can dissociate to give a stable carbocation this is the rate determining step. The subsequent reaction between the planar sp^2-hybridized carbocation and the nucleophile is very fast. The slow step only depends upon the alkyl halide and overall the reaction is called an S_N1 process (substitution, nucleophilic, unimolecular) (eqn 5.1). This is a typical reaction mechanism of tertiary halides, where the corresponding tertiary carbocation derives stability through hyperconjugation.

$$(CH_3)_3C\text{-}Cl \quad \underset{slow}{\overset{-Cl^-}{\rightleftharpoons}} \quad CH_3 \overset{+}{\underset{CH_3}{\diagdown}} CH_3 \quad \underset{fast}{\overset{Nu^-}{\rightleftharpoons}} \quad (CH_3)_3C\text{-}Nu \qquad (5.1)$$

Simple primary carbocations are too unstable to be formed in this way, and here the nucleophile approaches the alkyl halide so that the angle between it and the departing halogen is 180°. A transition state develops in which the nucleophile and the halide ion are partially bonded to the alkyl residue. The reaction is thus bimolecular and the rate depends on both the concentrations of the nucleophile and the alkyl halide: it is designated an S_N2 process (substitution, nucleophilic, unimolecular) (eqn 5.2).

$$Nu^- + CH_3CH_2\text{-}Cl \underset{slow}{\rightleftharpoons} \left[Nu \cdots \overset{H \quad CH_3}{\underset{H}{|}} \cdots Cl \right]^- \rightleftharpoons Nu\text{-}CH_2CH_3 + Cl^- \qquad (5.2)$$

In the case of chlorobenzene, resonance between the lone pair electrons of the halogen atom and the π system generates some double bond character (Fig. 5.1), and thus extra strength, to the carbon–halogen bond (C–Cl bond lengths are 170 pm for chlorobenzene and 178 pm for ^{t}butyl chloride). Further, the positive charge of the corresponding carbocation formed by the loss of the chloride anion would be accommodated in an sp^2 orbital orientated at right angles to the π system, thereby preventing effective mutual overlap and charge delocalization. For these reasons the S_N1 reaction is not observed.

Whereas insufficient energy is available to allow dissociation of the C–Cl bond and the formation of the benzene carbocation, such species may be produced during the decomposition of diazonium salts. Here the elimination of a nitrogen molecule is a highly exothermic event providing the driving force for the reaction (see p.59).

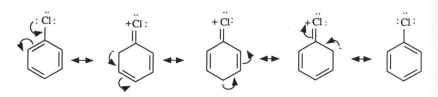

Fig. 5.1 Resonance in chlorobenzene

The typical S_N2 process is equally disfavoured, since apart from repulsion between the nucleophile and the π system, the geometry of the ring prevents the development of the usual S_N2 transition state where the nucleophile and the leaving group are aligned in a linear fashion. Despite this, nucleophilic aromatic substitution reactions showing second order kinetics do occur, but only for arenes containing strongly electron withdrawing groups in the *ortho* and/or *para* positions. An addition/elimination sequence is followed and 2,4,6-trinitrochlorobenzene, for example, can be reacted with sodium methoxide to form 2,4,6-trinitromethoxy-benzene, (eqn 5.3). The effect of the nitro groups is firstly to reduce the electron density of the π system and allow the nucleophile (MeO⁻) to approach the ring and form a negatively charged sigma complex (a Meisenheimer complex). This is then stabilized through delocalization of the negative charge with the nitro groups. The forward reaction eliminates chloride ion and reforms the aromatic ring. Where the nucleophile and the leaving group are chemically similar, mixtures form. For example, if 2,4,6-trinitromethoxybenzene is crystallized from ethanol a mixture of starting material and 2,4,6-trinitroethoxybenzene is produced. This illustrates just how easily a Meisenheimer complex can form: here the activational effect is maximized since all three nitro groups participate in the delocalization of the negative charge (eqn 5.4).

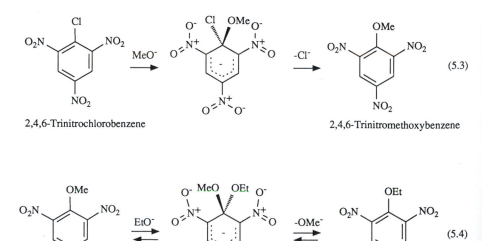

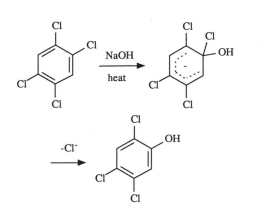

It is not always necessary to delocalize the negative charge of the Meisenheimer complex through conjugation with an unsaturated group such as NO$_2$. The presence of several strongly electronegative ions may achieve the same result through inductive effects: thus 1,2,4,5-tetrachlorobenzene reacts with hot sodium hydroxide solution to afford 2,4,5-trichlorophenol (eqn 5.5).

$$(5.5)$$

5.2 Aryne formation

Following these comments it might be supposed that a simple aryl halide like chlorobenzene would not react with nucleophiles. However, if chlorobenzene is heated at 350 °C with sodium hydroxide under high pressure, sodium phenate is produced. Similarly if bromobenzene is treated with potassium amide, aniline is produced. Potassium amide is a very strong base and if bromobenzene which has been isotopically labelled at C-1 is used, it can be shown that the label is almost equally divided between C-1 and C-2 in the product (eqn 5.6). *A direct replacement of the halogen atom does not occur*, instead a highly strained and reactive molecule named benzyne is formed. The role of the strong base is thus to eliminate hydrogen bromide (as ammonia and bromide ion). Amide anion then attacks benzyne, prior to protonation and the formation of aniline. Overall this is an elimination–addition process, and since benzyne is symmetrical the addition of ammonia cannot be selective.

The first step of eqn 5.6 is not concerted, but is an example of an E1$_c$B reaction (Elimination from a conjugate Base). Deprotonation to an intermediate carbanion is fast and reversible but loss of bromide ion from the intermediate is slow and rate limiting.

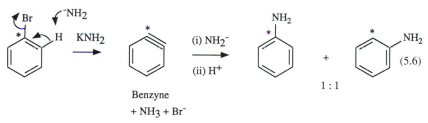

Benzyne is the parent of a series of related molecules, collectively known as arynes, which can be synthesized from a variety of starting materials. For example, arynes are formed by the action of lithium or magnesium on 1,2-dihalobenzenes (eqn 5.7). Benzyne is also considered to be generated in the reaction of chlorobenzene with sodium hydroxide,but since the latter reagent is a weaker base than potassium amide much more vigorous conditions are required (eqn 5.8).

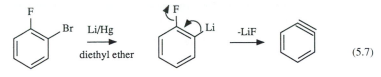

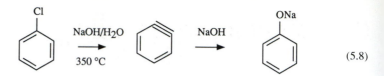

$$(5.8)$$

Benzyne reacts with itself, in the absence of a suitable nucleophile, to give biphenylene (eqn 5.9). Furthermore, it combines with dienes (as a dienophile) to give cyclo adducts. For example, with anthracene (see p. 80) benzyne yields triptycene (eqn 5.10).

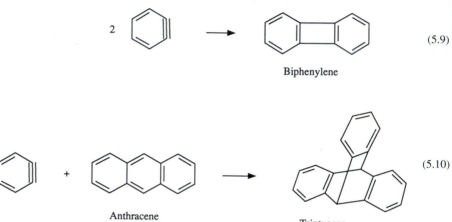

$$(5.9)$$

$$(5.10)$$

6 Aromatic compounds

6.1 Arylalkanes

Methylbenzene is better known as toluene, and the dimethylbenzenes as *o-*, *m-*, and *p*-xylenes, depending on the relative orientations of the two methyl groups. 1,3,5-Methylbenzene is called mesitylene, and 1,2,4,5-tetramethylbenzene, durene. Previously benzene, toluene, and the xylenes were obtained almost exclusively from the light oil fraction of coal tar, but in modern times a major source is petroleum, via the aromatization of other hydrocarbons. Laboratory syntheses are usually achieved through direct alkylation of arenes by the Friedel–Crafts method (p. 14), although it is also possible to reduce carbonyl groups directly bonded to the ring by the Wolff–Kishner procedure (see p. 72) or the Clemmensen reaction, just as in the aliphatic series.

Alkenylbenzenes are normally synthesized through the dehydration of arylalkanols, or by the dehydrohalogenation of arylalkyl halides (Fig. 6.1). The best known alkenylarene, ethenylbenzene (styrene), is obtained commercially from benzene in two steps: (i) alkylation with ethene in the presence of an acid or a Lewis acid, and (ii) dehydrogenation of the product, ethylbenzene, by heating over a catalyst at 600 °C (eqn 6.1).

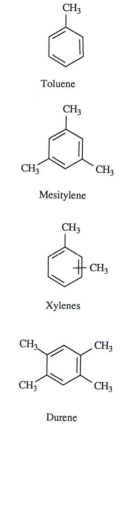

Toluene

Mesitylene

Xylenes

Durene

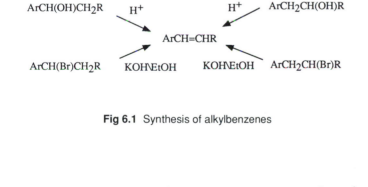

Fig 6.1 Synthesis of alkylbenzenes

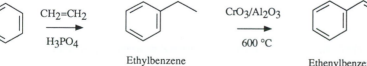

(6.1)

Ethylbenzene

Ethenylbenzene
(styrene)

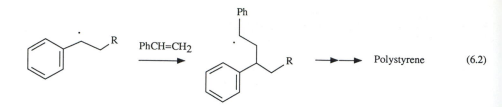

$$\text{(6.2)}$$

Styrene undergoes radical initiated polymerization with itself, to form polystyrene (eqn 6.2), or with other molecules to form copolymers; both processes have important commercial applications. First, attack of an initiating radical (R·) is favoured at the β-carbon atom of the side chain to give the *benzylic* radical (i.e. the radical centre is situated on the carbon next to the benzene ring) which, although resonance stabilized, enters readily into chain reactions utilizing the α carbon as the centre for the chain extending process. Electrophiles (R⁺) similarly attack the β position of styrene and thus afford benzylic cations, which are also resonance hybrids in which the positive charge is delocalized by the π system of the benzene nucleus (Fig. 6.2). In the presence of a nucleophile (X⁻) the carbocation is attacked at the α carbon (eqn 6.3); thus the overall process is an addition to the double bond which, in the case of styrene itself, follows the Markownikoff rule (i.e. the attack of the electrophile proceeds to give the 'better' carbocation, prior to the addition of the nucleophile).

Substituents in the benzene ring may influence the mode of reaction. For example, strongly electron withdrawing groups at the *ortho* or *para* sites will destabilize a cationic centre at the benzylic position and thus favour anti-Markownikoff addition.

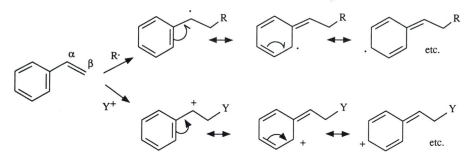

Fig. 6.2 Resonance in benzylic radicals and cations

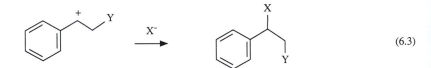

$$\text{(6.3)}$$

It is important to note that, although the carbocation is resonance stabilized, the major contributor has the charge located in the side chain thereby retaining an aromatic sextet of electrons. For this reason reaction with the nucleophile occurs in the side chain and not in the nucleus. Hydrogen bromide, reacts with styrene and gives 1-bromoethylbenzene (eqn 6.4a), but if a peroxide initiator is present bromine atoms are produced, and these react with styrene to form the benzylic radical. The radical abstracts a hydrogen atom from hydrogen bromide to give 2-bromoethylbenzene and another bromine atom. A chain reaction is thus established (eqn 6.4b).

This is an example of the Kharasch peroxide effect which reverses the normal (ionic) regioselectivity of the addition of hydrogen bromide to alkenes.

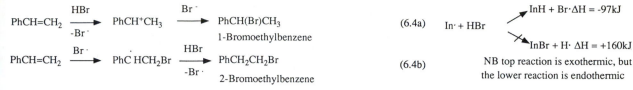

$$PhCH=CH_2 \xrightarrow[-Br^-]{HBr} PhCH^+CH_3 \xrightarrow{Br^-} PhCH(Br)CH_3 \qquad (6.4a)$$
1-Bromoethylbenzene

$$PhCH=CH_2 \xrightarrow{Br \cdot} PhC \overset{\cdot}{H}CH_2Br \xrightarrow[-Br\cdot]{HBr} PhCH_2CH_2Br \qquad (6.4b)$$
2-Bromoethylbenzene

$$In \cdot + HBr \begin{array}{l} \nearrow InH + Br \cdot \Delta H = -97kJ \\ \searrow InBr + H \cdot \Delta H = +160kJ \end{array}$$

NB top reaction is exothermic, but the lower reaction is endothermic

Styrene adds halogens across the double bond and can be hydrogenated under relatively mild conditions to give ethylbenzene (Fig. 6.3); it also reacts with peroxides to afford the corresponding oxirane, styrene oxide. This molecule can be ring opened with nucleophiles, which attack at the less hindered β position. If the nucleophile is phenyllithium the product is 1,2-diphenylethanol, dehydration of which leads to *trans*-stilbene (eqn 6.5).

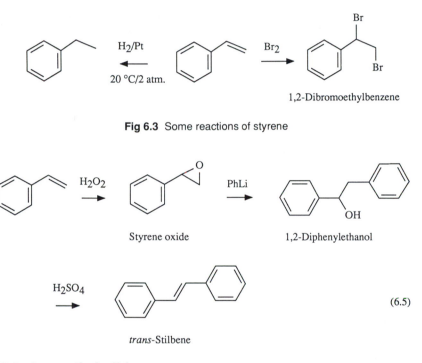

1,2-Dibromoethylbenzene

Fig 6.3 Some reactions of styrene

Styrene oxide   1,2-Diphenylethanol

(6.5)

trans-Stilbene

6.2 Aromatic halides

6.2.1 Nomenclature

Aryl halides (halobenzenes) are compounds in which the halogen atom is directly bonded to the aromatic ring. Benzyl halides (phenylmethyl halides) have a single halogen atom attached to the α carbon atom of an alkyl side chain (i.e. the carbon atom adjacent to the aromatic nucleus). The terms benzal and benzo refer to phenylmethyl di- and tri-halo compounds respectively (see p. 34). The position of halogen atoms at other positions in the side chains of alkyl benzenes should be denoted by either a number or a Greek letter.

6.2.2 Synthesis

The usual route to aryl halides is through the direct halogenation of the arenes in the presence of a Lewis acid (see p. 14). It is also possible to use the diazonium salts as starting materials (see p. 59).

Arylalkyl halides (X = Cl or Br) are obtained either from the corresponding alcohols by reaction with the appropriate phosphorus halides, or halogen acids (eqn 6.6), or by the addition of halogen acids to arylalkenes (see above).

If the hydroxyl group is benzylic, the corresponding carbocation is stabilized by resonance and is easily formed. This species may then eliminate a proton and give the corresponding styrene, rather than a benzyl halide.

$$\mathrm{Ar(CH_2)_nCH(OH)R} \xrightarrow[\mathrm{-YOH}]{\mathrm{YX}} \mathrm{Ar(CH)_nCH(X)R} \tag{6.6}$$

(YX = POCl$_3$, PBr$_3$, HBr, etc. n = 0, 1, 2...)

Benzyl chlorides are available through the chloromethylation of arenes with formaldehyde and hydrogen chloride. Direct chlorination of toluene in sunlight, or treatment with sulphuryl chloride in the presence of dibenzoylperoxide as initiator, gives, in sequence, benzyl chloride, benzal chloride, and benzotrichloride (eqn 6.7a, b). This reaction is similar to the radical chain chlorination of methane, and is facilitated because the benzylic radicals involved are stabilized through the delocalization of their unpaired electron with the pi system of the aromatic nucleus.

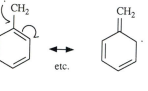

etc.

$$\mathrm{Cl_2} \xrightarrow{h\upsilon} 2\mathrm{Cl}\cdot \tag{6.7a}$$

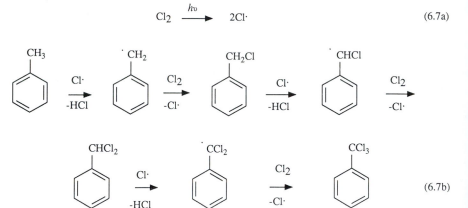

$$\tag{6.7b}$$

Due to the high reactivity of chlorine atoms, higher homologues of toluene are not attacked selectively. In the case of ethylbenzene, for example, a mixture of 1-chloro- and 2-chloroethylbenzene is produced (eqn 6.8a). Bromine atoms are less reactive so that more radical character develops in the transition state. Therefore the bromination reaction is more regioselective giving only 1-bromoethylbenzene (eqn 6.8b).

$$\mathrm{PhCH_2CH_3 + Cl_2} \xrightarrow{h\upsilon} \mathrm{PhCH(Cl)CH_3 + PhCH_2CH_2Cl} \tag{6.8a}$$
$$ 56\% \qquad\quad 44\%$$

$$\mathrm{PhCH_2CH_3 + Br_2} \xrightarrow{h\upsilon} \mathrm{PhCH(Br)CH_3} \tag{6.8b}$$

6.2.3 Reactions

When reacted with lithium and magnesium, aryl bromides and iodides give aryl lithiums and aryl Grignard reagents, respectively (Fig. 6.4). (Chlorides also react, but the reactions are slower and side products are formed, unless the solvent is tetrahydrofuran; see p. 36.) Aryl Grignard and aryl lithium reagents undergo similar reactions to those shown by their aliphatic counterparts, combining with a range of electrophiles (carbonyl compounds, nitriles, orthoesters etc.) to give a wide selection of important derivatives (Fig. 6.5).

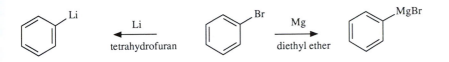

Fig. 6.4 Reactions of aryl bromides with lithium and magnesium

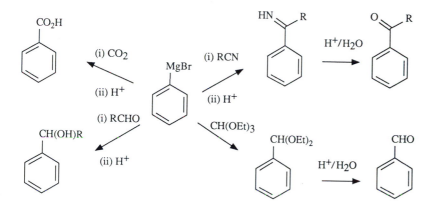

Fig. 6.5 Reactions of aryl Grignard and aryllithium reagents

The Grignard and lithium reagents of arylalkyl halides are prepared and react similarly, but benzylmagnesium halides are often contaminated with bibenzyls which are formed as byproducts. It should be noted that Grignard reagents are bases capable of forming enolate anions from carbonyl compounds with α hydrogen atoms. In such cases attack at the carbonyl centre may be superseded by aldol type reactions where the enolate combines with a second molecule of the carbonyl compound (eqn 6.9).

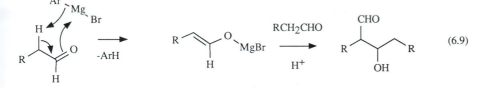

(6.9)

The metallations of both aryl and alkyl halides (RX) proceed through a series of one-electron transfer steps, but for organolithiums and Grignard reagents there is little evidence to show that the bond between the metal and the organic group is anything other than covalent (eqn 6.10).

$$RX + Mg \longrightarrow RX^{\cdot -}Mg^{\cdot +} \longrightarrow R^{\cdot} + Mg^{\cdot +}X^{-} \longrightarrow RMgX \qquad (6.10)$$

There has been much discussion of the nature of the actual species present in solution and it has been shown that aggregates tend to form. The nature of the solvent is important. For example, arylmagnesium halides are in equilibrium with diarylmagnesiums and the magnesium halide (eqn 6.11). In diethyl ether the arylmagnesium halide is preferred, whereas in tetrahydrofuran its predominance is less and in some cases more diarylmagnesium is present. This can influence the reactivity of the reagent.

$$2ArMgX \; \rightleftharpoons \; Ar_2Mg + MgX_2 \; \rightleftharpoons \; Ar_2Mg.MgX_2 \qquad (6.11)$$

Complexes nominally formulated as $Ar_2Mg.MgX_2$ may also be present — the Schlenk equilibrium.

Other metals such as sodium and potassium can replace magnesium and lithium. The addition of sodium to a mixture of an aryl halide and an alkyl halide (RX) produces an alkyl arene (Würtz–Fittig reaction) (eqn 6.12). Yields are generally poor.

$$ArX + RX \quad \xrightarrow[-NaX]{Na} \quad ArR \qquad (6.12)$$

Aryl iodides couple with lithium copper reagents (R_2CuI) to afford substituted aryls (R = primary and, sometimes, secondary alkyl, allyl, vinyl, and benzyl) (eqn 6.13). Arylalkyls are also obtained when arylmagnesium bromides are reacted with primary alkyl halides in the presence of cuprous ions (eqn 6.14). Aryl iodides when heated with copper afford biaryls: this is the basis of Ullmann reaction. Aryl copper intermediates [ArCu] are initially formed, but the precise structures of these compounds are uncertain (eqn 6.15).

The first step in the Ullmann reaction may be radical in nature, but the second step is possibly a nucleophilic attack by [ArCu] on ArI.

$$ArI + R_2CuI \quad \longrightarrow \quad ArR \qquad (6.13)$$

$$ArMgBr + RBr \quad \xrightarrow[Et_2O]{Cu^+} \quad ArR \qquad (6.14)$$

$$ArI \;\rightarrow\; [ArCu] \xrightarrow{\;ArI\;} Ar\text{–}Ar \qquad (6.15)$$

The nucleophilic substitution reactions of aryl halides in which the halogen atom is directly bonded to the ring have already been discussed (see p. 27). When the halogen is in a side chain the compound behaves as an alkyl halide, except when the carbon atom bearing the halogen is bonded to the ring, in which case the corresponding carbocation is stabilized by resonance. Thus even though the benzyl cation is primary, the energy requirement for its formation is relatively low and benzyl halides react with nucleophiles by the S_N1 mechanism.

β-Arylethyl halides (phenethyl halides) react with nucleophiles by two mechanisms: one is a straight forward S_N2 process, whereas the other involves neighbouring group participation (anchimeric assistance) from the π electrons of the aryl ring. These 'push out' the halide ion to form a bridged carbocation called a *phenonium ion* which is then attacked by the nucleophile (Fig. 6.6).

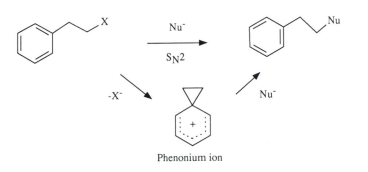

Phenonium ion

Fig. 6.6 Reaction of β-arylethyl halides with nucleophiles

In the absence of suitable nucleophiles, phenonium ions can be analysed by NMR spectroscopy: they show no aromatic character and thus resemble arenium ions. Such species are not solely restricted to β-arylethyl halides, but are recognized as intermediates in similar reactions of other β-arylethyl derivatives, such as sulphonates, acetates etc., where there is the potential for a 'good leaving group' to form. The involvement of a phenonium ion can have a significant rate enhancing effect.

The triphenylmethyl cation and the triphenylmethyl radical are both easily produced, and each is further stabilized by resonance with the benzene nuclei. A solution of triphenylmethyl chloride in sulphur dioxide for example, conducts electricity due to heterolytic dissociation of the C–Cl bond (eqn 6.16), whereas if triphenylmethyl chloride is reacted with silver in the non-polar solvent benzene a yellow solution forms, due to homogenous fission of the bond. The triphenylmethyl radical combines with oxygen to afford the peroxide, or reacts with itself to yield an adduct (Fig. 6.7). At one time this adduct was considered to be hexaphenylethane, but such a compound does not exist because of the considerable steric overcrowding inherent in this structure.

Neither the triphenylmethyl carbocation nor the triphenylmethyl radical are planar: they adopt a propeller geometry to minimize non-bonded interactions between the *ortho* protons.

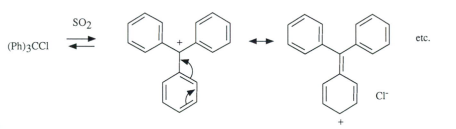

(Ph)₃CCl

etc.

(6.16)

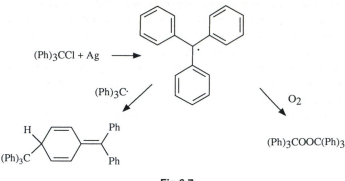

Fig 6.7

6.3 Aromatic alcohols

6.3.1 Nomenclature

Compounds in which the hydroxyl group is bonded directly to the aromatic nucleus are called phenols. They have different properties to the aromatic alcohols (arylalkanols), in which the hydroxyl group is attached to an atom of an alkyl side chain. The phenols are considered below in Section 6.9. Benzyl alcohol (phenylmethanol) is the most familiar aromatic alcohol.

6.3.2 Synthesis

Benzyl alcohol is synthesized by the hydrolysis of benzyl chloride, or by the reduction of benzaldehyde (eqn 6.17). Other members of the series are available by similar reactions or through the action of aryl or alkyl Grignard reagents on carbonyl compounds (eqn 6.18a, b).

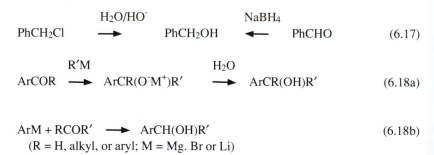

6.3.3 Reactions

In general the reactions of arylalkyl alcohols are similar to those of alkanols, and their ethers, esters etc. are produced by standard methods. Oxidation of primary arylalkanols affords first the aldehyde and then the corresponding acid, whereas secondary arylalkanols give ketones. When the hydroxyl group is in the benzylic position, protonation is followed by rapid loss of water to give the benzylic carbocation which is resonance stabilized. Further reaction may occur with available nucleophiles; if these are absent polymerization takes place (see p. 65).

Benzyl ethers are used in synthesis as protection groups because they are readily cleaved either by treatment with acid or by hydrogenolysis over a palladium catalyst. The acid promoted reaction involves the formation of the benzyl cation (eqn 6.19), whereas hydrogenolysis depends upon the intermediacy of palladated species.

4-Methoxybenzyl ethers are particularly easily cleaved.

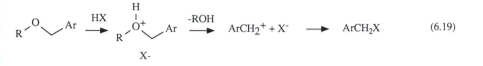

$$\text{(6.19)}$$

6.4 Aromatic carbonyl compounds

6.4.1 Araldehydes

Benzaldehyde is the parent of a series of aromatic aldehydes. Resonance between the formyl group and the ring ensures that the latter is deactivated towards electrophilic attack (Fig. 6.8). In addition, the reactivity of the formyl group towards nucleophiles is reduced compared to that of alkanals.

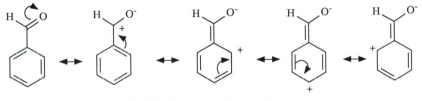

Fig. 6.8 Resonance in benzaldehyde

6.4.1.1 Synthesis
Benzaldehyde is synthesized by the Gatterman–Koch reaction (see p. 16), or through hydrolysis of benzal chloride in the presence of alkali (eqn 6.20).

Although benzene is insufficiently active to undergo the Vilsmeier formylation reaction (see p. 16), this reaction is often used to prepare other more electron rich araldehydes. Partial reduction of acid halides, nitriles, and esters can also be employed.

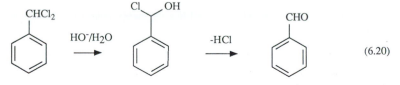

$$\text{(6.20)}$$

6.4.1.2 Reactions

Aromatic aldehydes can be oxidized to acids and reduced to alcohols. They also react, although less well, in a similar manner to their aliphatic counterparts with a wide range of nucleophiles. For example, they combine with primary aromatic amines (ArNH$_2$), arylhydrazines (ArNHNH$_2$), semicarbazide (NH$_2$NHCONH$_2$), and hydroxylamine (NH$_2$OH) by a similar pathway. Initial nucleophilic attack by the lone pair electrons of the nitrogen derivative at the carbon atom of the carbonyl group is followed by protonation and the elimination of a molecule of water to give imines (Schiff's bases), hydrazones, semicarbazones, and oximes respectively (eqn 6.21).

Reaction with sodium bisulphite yields bisulphite compounds, from which cyanohydrins can be obtained by treatment with sodium cyanide (eqn 6.22). The direct formation of cyanohydrins is complicated by the propensity of aromatic aldehydes to form benzoins. Thus the reaction of benzaldehyde with sodium cyanide in ethanol solution gives rise to benzoin itself. The initial step is identical with that of cyanohydrin formation, but in ethanol protonation is slower than internal charge redistribution and the carbanion so formed now reacts with a second molecule of benzaldehyde to give the hydroxy ketone (Fig. 6.9).

The conversion of the bisulphite complex into the cyanohydrin probably involves a direct substitution reaction.

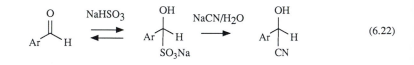

(6.21)

(R=Ar, NHR, NHCONH$_2$, or OH)

The benzoin reaction relies upon CN⁻ being nucleophilic and capable of stabilizing α carbanions, as well as acting as a relatively good leaving group.

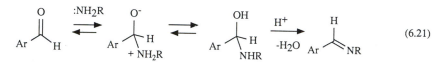

(6.22)

Certain araldehydes, such as *p-N,N*-dimethylamino- and *p*-nitrobenzaldehydes, fail to undergo the benzoin reaction. For the first example, this is because the carbonyl group is insufficiently nucleophilic (through conjugation with the lone pair electrons of the amino substituent); for the second, the initially formed carbanion is stabilized through resonance with the nitro group thus becoming unreactive towards another molecule of aldehyde.

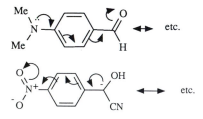

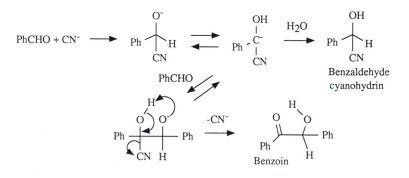

Fig. 6.9 Reaction of benzaldehyde with sodium cyanide

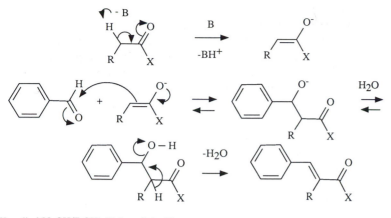

X = alkyl(NaOH/EtOH) Claisen–Schmidt
X = O-alkyl(NaOEt) Claisen
X = OCOR(NaOAc)/180 °C) Perkin
X = OEt; R = CO$_2$Et(piperidine/100 °C)Knoevenagel

Fig. 6.10 Reactions of aromatic aldehydes

Aromatic aldehydes cannot be enolized since they lack a hydrogen atom α to the formyl group. In this respect they resemble formaldehyde. None the less, they react readily with carbanions formed from suitable aldehydes, ketones, esters, and anhydrides. Again, initial nucleophilic addition is followed by the elimination of water so that α,β-unsaturated products form. By eliminating a molecule of water an enone unit is formed, the conjugation of which is extended by the benzene π system. As a result the products acquire considerable thermodynamic stability. Most of these reactions have long histories and they are named after their discoverers (Fig. 6.10).

In the Perkin reaction (eqn 6.23) a cyclic intermediate facilitates the 'dehydration process', and it is instructive to note that the severe reaction conditions are a reflection of the weak basicity of the acetate anion which is traditionally used to deprotonate the anhydride. If sodium hydride is employed as base, the reaction occurs at -5 °C.

The enones are frequently less soluble than the first formed hydroxy carbonyl compounds, causing them to separate out from the reaction mixture.

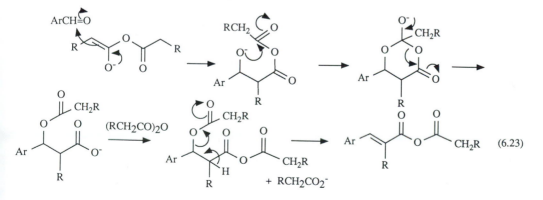

(6.23)

The transfer of 'hydride ion' is facilitated through intermolecular association between the initially formed sodium salt and a second molecule of the aldehyde:chelation control.

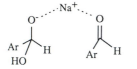

The Cannizzaro reaction (eqn 6.24) is an interesting example of a mutual oxidation–reduction procedure, in which an aromatic aldehyde is reacted with a concentrated solution of sodium hydroxide. The anion produced transfers a hydride ion or its equivalent to a second molecule of the aldehyde. The reaction can be made synthetically useful if an equimolar mixture of an araldehyde and formaldehyde is used (eqn 6.25). Since formaldehyde is more reactive towards nucleophiles than the araldehyde (see p. 39), it is attacked by hydroxide ion faster. Thus it is oxidized to formic acid and the aromatic aldehyde is converted into the corresponding alcohol.

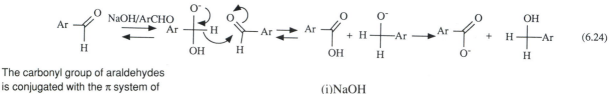

$$(6.24)$$

The carbonyl group of araldehydes is conjugated with the π system of the ring, which has the effect of lowering nucleophilicity at the carbon atom, unless there are strongly electron withdrawing groups at the *o*- and/or *p*-positions.

$$\text{HCHO} + \text{ArCHO} \quad \xrightarrow[\text{(ii)H}^+]{\text{(i)NaOH}} \quad \text{HCO}_2\text{H} + \text{ArCH}_2\text{OH} \qquad (6.25)$$

6.4.2 Aromatic ketones

Arylalkyl and arylaryl ketones are normally synthesized by Friedel–Crafts acylation reactions, or if the ring bears strongly electron donating substituents by the Hoesch reaction (see p. 17).

They react in similar ways to aromatic aldehydes, although less readily since the products are more sterically crowded, and when the carbonyl group is flanked by two aromatic nuclei, the possibilities for resonance are enhanced and the electrophilicity of the carbonyl carbon atom is reduced. The effect of resonance can be readily assessed from the infra red carbonyl stretching frequency which gives a direct reflection of single/double bond character: the lower the wave number the greater the ionic character of the bond and vice versa.

Oxidation of benzoin with nitric acid gives the diketone benzil, which when treated with concentrated alkali rearranges to benzylic acid (eqn 6.26).

Acetophenone

μ_{max} 1690cm^{-1}

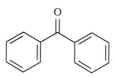

Acetone

μ_{max} 1720cm^{-1}

Benzophenone

μ_{max} 1640cm^{-1}

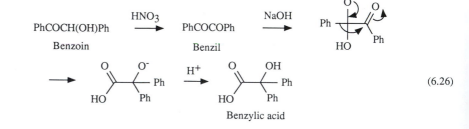

$$(6.26)$$

Benzylic acid

6.5 Aromatic acids, esters, and amides

6.5.1 Synthesis of acids

Benzoic acid ($PhCO_2H$) is synthesized by the hydrolysis of benzotrichloride or benzonitrile; the oxidation of toluene, benzyl alcohol or benzaldehyde; or the reaction of phenylmagnesium bromide with carbon dioxide (Fig. 6.11). Other aromatic acids are obtained by similar methods.

$$PhCCl_3 \xrightarrow[H_2O]{NaOH} PhCO_2H \xleftarrow{NaOH} PhCN$$

$$PhCH_3 \xrightarrow[Co(OAc)_2]{O_2} PhCO_2H \text{ (commercial source)}$$

$$PhCH_2OH \xrightarrow{KMnO_4} PhCHO \xrightarrow{[O]} PhCO_2H$$

$$PhMgBr \xrightarrow{CO_2} PhCO_2MgBr \xrightarrow{H^+} PhCO_2H$$

Fig. 6.11 Synthesis of benzoic acid

6.5.2 Acidity

The acidity of a carboxylic acid is a measure of the position of the equilibrium between the free acid and the carboxylate anion (eqn 6.27). The stability of the carboxylate anion is influenced by both inductive and resonance effects. Benzoic acid (pK_a 4.2) is somewhat more acidic than acetic acid (pK_a 4.8) because the carboxylic acid group is bonded to an sp^2- rather than to an sp^3-hybridized carbon atom.

The electronegativity of carbon atoms is in the order $sp > sp^2 > sp^3$.

(6.27)

The acidities of some benzoic acids are recorded in Table 6.1. The results are not totally predictable.

Table 6.1. The acidity of prominent benzoic acids

Substituent	pK_a		
	o-	*m-*	*p-*
F	3.3	3.9	4.1
Cl	2.9	3.8	4.0
Br	2.9	3.8	4.0
HO	3.0	4.1	4.6
MeO	4.1	4.1	4.5
NO_2	2.2	3.5	3.4

Note: in neither case can the negative charge of the carboxylate ion be conjugated directly with the π system of the ring. A partial positive charge at the *ipso* site, however, assists in reducing the full negative charge of the carboxylate groups.

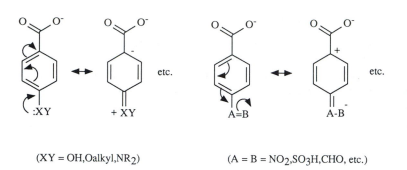

(XY = OH,Oalkyl,NR$_2$) (A = B = NO$_2$,SO$_3$H,CHO, etc.)

Fig. 6.12 Resonance effects of electron donating groups (XY) and electron withdrawing groups (A=B) at the *para* position in the carboxylate anions

Electron donating groups, such as hydroxy, alkoxy, and amino, at the *o* and *p* positions of benzoic acid reduce acidity through the resonance effect (Fig. 6.12). This increases the electron density of the ring and destabilizes the conjugate anion, even though the inductive effect works in the opposite direction and is greatest at the *ortho* and least at the *para* position. The situation is complicated, however, by the fact that the *o* substituents interfere sterically with the carboxyl group, reducing its coplanarity with the π system of the ring. The steric factor applies equally to benzoic acids bearing electron withdrawing groups at the *o* position. Such groups bonded to either the *o* or *p* sites, through resonance (see Fig. 6.12), allow the development of positive charge adjacent to the carboxyl group. This has the opposite effect to that of electron donating groups and increased acidity is to be expected since this reinforces the inductive effect of the substituent group and also stabilizes the anion.

6.5.3 Reactions of acids

Aromatic acids can be esterified by reaction with alcohols and sulphuric acid, or with hydrogen chloride (eqn 6.28). Acid chlorides are obtained through the action of phosphorus trichloride or thionyl chloride (eqn 6.29).

$$\text{ArCO}_2\text{H} + \text{ROH} \xrightarrow{\text{HX}} \text{ArCO}_2\text{R} + \text{H}_2\text{O} \qquad (6.28)$$

$$\text{ArCO}_2\text{H} \xrightarrow{\text{PCl}_3(\text{orSOCl}_2)} \text{ArCOCl} \qquad (6.29)$$

The term anilide also includes *N*-acylarylamines.

Arylacid halides react with phenols to yield aryl benzoates, and with arylamines (anilines) in the presence of sodium hydroxide or other bases to give *N*-aroylarylamines (anilides). These are examples of the Schotten–Baumann procedure (eqn 6.30a, b).

$$ArCOCl + Ar'OH \xrightarrow{\text{NaOH}} ArCO_2Ar' + HCl \qquad (6.30a)$$

$$ArCOCl + Ar'NH_2 \xrightarrow{\text{NaOH}} ArCONHAr' + HCl \qquad (6.30b)$$

Aliphatic amines can be used in place of arylamines and sodium hydroxide can be replaced by other bases including sodium acetate or the tertiary amine pyridine. Benzamide is obtained from the reaction of benzoyl chloride and ammonia (eqn 6.31). It is a neutral compound, the lone pair electrons on the nitrogen atom being involved in resonance with the carbonyl group.

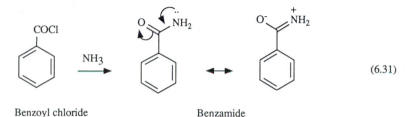

Benzoyl chloride   Benzamide

6.5.4 Directed *ortho* lithiation of benzamides

Tertiary benzamides (ArCONR$_2$) can be lithiated at the *ortho* position by reaction with an alkyl lithium (frequently ^{s}butyl or ^{t}butyl) in a non-polar solvent such as tetrahydrofuran containing tetramethylethylenediamine (TMEDA). The reaction involves initial coordination of the alkyl lithium and the amide group, followed by deprotonation and the formation of an *ortho*-chelated complex. Such complexes react regioselectively with a wide range of electrophiles yielding *ortho*-substituted benzamides (eqn 6.32).

Alkyl lithiums form aggregates in organic solvents, and TMEDA acts to break these down into monomers or dimers thereby significantly increasing their basicity.

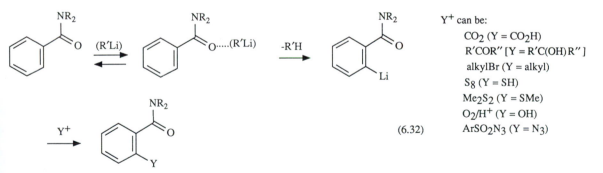

Y^+ can be:

CO_2 (Y = CO_2H)
R'COR″ [Y = R'C(OH)R″]
alkylBr (Y = alkyl)
S_8 (Y = SH)
Me_2S_2 (Y = SMe)
O_2/H^+ (Y = OH)
$ArSO_2N_3$ (Y = N_3)

Substituents may be present in the aromatic rings of the benzamides. When these are *meta*-alkoxy there is a strong preference for electrophilic substitution at the 2-position (eqn 6.33), probably because (a) the inductive effect of the oxygen atom increases the acidity of the *ortho* protons, and (b) the lone pair electrons on the alkoxy group participate in coordination to the lithium atom.

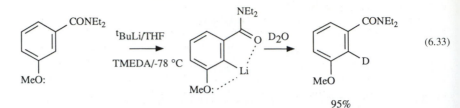

$$(6.33)$$

95%

Secondary benzamides can be used in place of tertiary benzamides, but in this case the compound is deprotonated first at the amidic position prior to *o* lithiation. Directed *ortho* lithiation has condiderable value in synthesis and the range of starting materials can be extended to include 4,5-dihydro-4,4-dimethyl-2-phenylisoxazoles (see Fig. 6.13) (which act as masked acids and are rather more easily hydrolysed than benzamides), Schiff's bases (arylimines), and the acetals of araldehydes (2-aryl-1,3-dioxolanes). After lithiation and reaction with an electrophile (Y^+), hydrolysis of either the Schiff's bases or the acetals yields *o*-substituted araldehydes (Fig. 6.14).

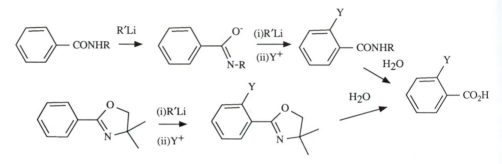

2-Aryl-4,5-dihydro-4,4-dimethyl-1,3-oxazoles

Fig. 6.13 *Ortho* lithiation of benzamides

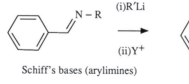

Schiff's bases (arylimines)

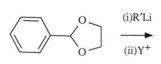

2-Aryl-1,3-dioxolanes

Fig. 6.14 Some synthetic applications of *ortho* lithiation

6.5.5 Dicarboxylic acids

Phthalic acid and terephthalic acid are important, the former in the manufacture of plasticizers and glyptal resins, and the latter in the production of condensation polymers, such as nylon and terylene. Commercially they are made from the oxidation of *o*-xylene (1,2-dimethylbenzene) or naphthalene, and *p*-xylene (1,4-dimethylbenzene), respectively (Fig. 6.15). On heating phthalic acid dehydrates to phthalic anhydride.

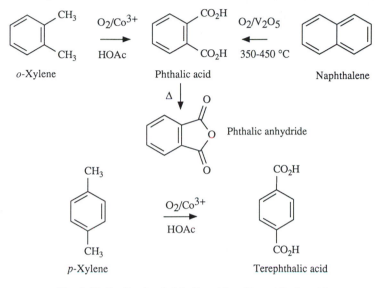

Fig. 6.15 Synthesis of phthalic acid and terephthalic acid

Phthalic anhydride when heated with ammonia gives phthalimide which when hydrolysed affords phthalamic acid (eqn 6.34). Phthalic acid anhydride can be used in Friedel–Crafts acylation reactions, and with benzene, for example, yields *o*-benzoylbenzoic acid — a precursor of anthraquinone (eqn 6.35).

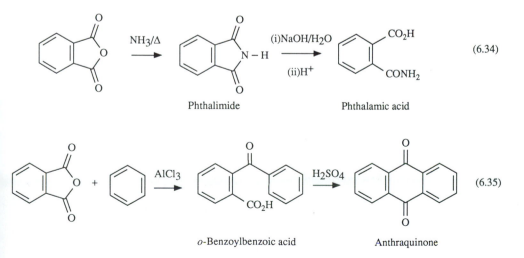

6.6 Aromatic sulphonic acids

Benzene sulphonic acid, obtained by the sulphonation of benzene with fuming sulphuric acid at 40 °C (see p. 13), is readily soluble in water but its salts are less so and they can be 'salted out'. Sulphonation reactions can be reversed at higher temperatures and this can be useful in synthesis, either for the separation of more easily sulphonated arenes from mixtures, or to protect an active site from the attack of other electrophiles. The sulphonic acid group is deactivating and directs to the *meta* position, thus benzene-1,3-disulphonic acid is obtained if benzene is sulphonated with fuming sulphuric acid at 90 °C (eqn 6.36).

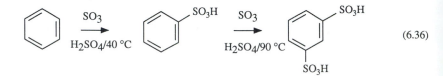

$$(6.36)$$

Sulphonyl chlorides are formed if sulphonic acids, or their sodium salts, are treated with phosphorus pentachloride or phosphorus oxychloride, or directly by the reaction of arenes with chlorosulphonic acid (eqn 6.37).

Strictly, the Schollen–Baumann reaction relates only to phenols and arylamines. When pyridine is used as the base it is possilbe that a *N*-sulphonyl pyridinium cation is formed. This acts as the attacking electrophile. Similar reactive cations, *N*-acylpyridiniums, are formed if pyridine is used in Schollen–Baumann reactions with acid chlorides etc. (see p. 44).

$$\underset{(R = H \text{ or } Na)}{ArSO_3R} \xrightarrow{PCl_5} ArSO_2Cl \xleftarrow{HSO_3Cl} ArH \qquad (6.37)$$

These compounds react with alcohols and phenols to afford sulphonates, and with ammonia and amines to give sulphonamides (eqn 6.38). Sodium hydroxide, or better a tertiary base such as pyridine, is used as a base in these further examples of the Schotten–Baumann reaction. Many sulphonamides have been made since they show antibacterial activity. Such compounds are now little used in human medicine, but have value in veterinary science.

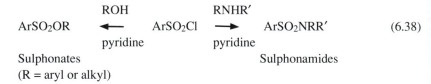

$$(6.38)$$

Sulphonates   Sulphonamides
(R = aryl or alkyl)

6.7 Nitro compounds and reduction products

6.7.1 Nitro compounds

Normally nitro compounds are prepared by the direct nitration of arenes, and occasionally by the oxidation of *C*-nitroso compounds. Diazonium salts can also be used as nitro group precursors when substituents in the parent arene direct nitration to other sites than that required.

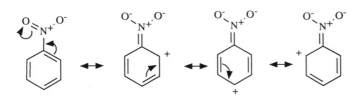

Fig. 6.16 Resonance in nitrobenzene

Resonance within arylnitro compounds deactivates the nucleus to electrophilic attack and ensures their polar nature (Fig. 6.16). Nitrobenzene is often used in reactions such as Friedel–Crafts acylations which need a solvent of high dielectric constant.

The nitro group also stabilizes a carbanion bonded to either the *o* or the *p* positions: thus *o*-nitrotoluene is deprotonated by treatment with base and the anion formed acts as a nucleophile towards carbonyl compounds such as aldehydes RCHO. The initial products readily dehydrate to give fully conjugated nitrostyrenes or nitrostilbenes (eqn 6.39).

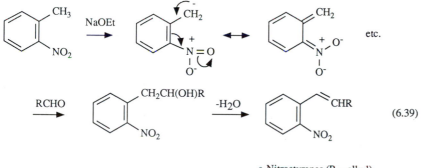

(6.39)

o-Nitrostyrenes (R = alkyl)
o-Nitrostilbenes (R = aryl)

6.7.2 Reduction of the nitro group

Nitrobenzene can be reduced to the primary aromatic amine, aniline (eqn 6.40) by hydrogenation over a catalyst such as platinum, or by employing a dissolving metal (normally zinc, tin, or iron) in hydrochloric or sulphuric acid. These methods are generally applicable to arylnitro compounds.

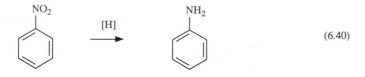

(6.40)

M − e $\longrightarrow$ M$^{\bullet+}$
M$^{\bullet+}$ − e $\longrightarrow$ M^{2+}
(M = Fe or Zn)

Electrochemical reduction at a mercury cathode is also effective, and when this is carried out in a protic medium and the cathode potential is carefully controlled, nitrobenzene itself gives firstly nitrosobenzene and then phenylhydroxylamine. The nechanism of the reduction has been studied and involves a series of electron addition–protonation steps (eqn 6.41). These compounds are also detected during the reduction of nitrobenzene with metals in acid solution and it is probable that the reaction operates through a similar pathway, the necessary electrons being released from the metal surface as it dissolves to form the metal cation.

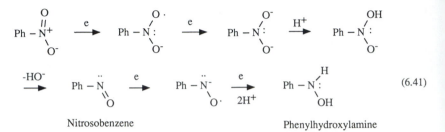

(6.41)

Nitrosobenzene   Phenylhydroxylamine

6.7.3 Azo compounds

The reduction of nitrobenzene to aniline under acidic conditions is a very complex process, and other compounds (azoxybenzene, azobenzene, and hydrazobenzene) are also formed. Normally these molecules do not survive, since they undergo further reduction to aniline, However, if the reduction conditions are varied, it is possible to isolate them: thus reduction of nitrobenzene with zinc in the presence of aqueous ammonium chloride gives phenylhydroxylamine, and reduction with arsenic trioxide in alkaline solution affords azoxybenzene. Hydrazobenzene can be synthesized by reducing nitrobenzene with hydrazine in alcoholic sodium hydroxide or in the presence of ruthenium on carbon as catalyst. An alternative method uses glucose and sodium hydroxide. Sodium dithionite can be employed for the individual reductions of azoxybenzene, azobenzene, and hydrazobenzene to aniline (Fig. 6.17).

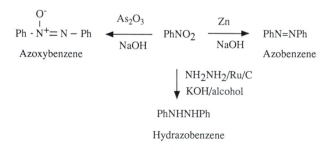

Fig. 6.17 Synthesis of azoxybenzene, hydrazobenzene, and azobenzene

$$PhNHOH \xrightarrow{HO^-} PhNHO^- \rightleftharpoons PhN^-OH \qquad (6.42a)$$

$$Ph-\overset{O}{\underset{..}{N}} \; PhN^-OH \longrightarrow Ph-\overset{OH}{\underset{|}{N}}-\overset{O^-}{\underset{|}{N}}-Ph \xrightarrow{-HO^-} PhN=N(O)Ph \qquad (6.42b)$$

$$PhNO + PhNH_2 \xrightarrow[-H_2O]{H^+} PhN=NPh \qquad (6.43)$$

Azoxybenzene is also prepared by reacting nitrosobenzene with phenylhydroxyylamine in alkaline solution (eqn 6.42 a, b), and azobenzene is formed by treating nitrosobenzene with aniline in the presence of acid (eqn 6.43).

Azoxybenzene is yellow in colour and azobenzene is a bright orange-red. Azobenzene is the parent of the class of compounds known as azo dyes (see p. 60). Note that hydrazobenzene is colourless, the single bond separating the nitrogen atoms preventing resonance between the π systems of the two benzene nuclei. Azobenzene can be oxidised to azoxybenzene by treatment with hydrogen peroxide. It normally exists in the near planar *E*-form, but irradiation with ultraviolet light causes it to rearrange to the *Z*-isomer (eqn 6.44). In this configuration steric hindrance between the two phenyl rings twists the molecule, and as a result the extended π system is distorted and the molecule is less highly coloured. It is also less stable and in solution reverts back to the *E*-isomer.

The colour is due to the highly conjugated nature of these compounds which lowers the energy separation between the highest occupied and the lowest unoccupied molecular orbitals so that absorption bands (π → π*) occur in the visible region of the electronic spectrum.

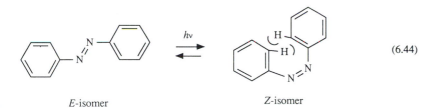

$$(6.44)$$

E-isomer   *Z*-isomer

Acid-catalysed rearrangement reactions are observed for both phenylhydroxylamine and hydrazobenzene. For example, although phenylhydroxylamine is reduced to aniline in acidic media, a competitive reaction is the Bamberger rearrangement to *p*-hydroxyaniline. A possible reaction mechanism is shown below in eqn 6.45.

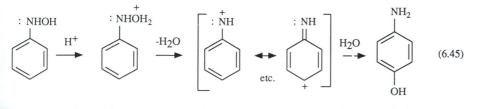

$$(6.45)$$

The remarkable benzidine rearrangement has been widely studied. Here hydrazobenzene (either as the mono- or di-protonated form) undergoes an intramolecular change to give 4,4′-diaminobiphenyl (benzidine), some 2,4′-diaminobiphenyl

Reactions of this type are known as sigmatropic rearrangements and are denoted by the number of atoms between the termini of the old and the new sigma bonds. The rearrangement to benzidine is thus an example of a [5,5]sigmatrophic shift.

(p-semidine), and minor amounts of 2,2′-diaminobiphenyl (o-semidine) and N-(2-and 4-aminophenyl)-anilines (eqn 6.46). The benzidine rearrangement reaction involves reorgnization of the electrons of the single bond between the two nitrogen atoms and the pi electrons to generate a new bond uniting two ring carbon atoms. One suggestion for the mechanism of the reaction under conditions of monoprotonation is that the N–N bond fragments to give a π complex. The two aryl units then rotate about one another so that the centres for the new sigma bonds of the various products are brought into close proximity.

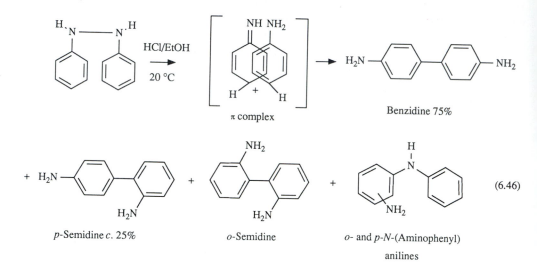

π complex

Benzidine 75%

p-Semidine *c*. 25%

o-Semidine

o- and *p*-*N*-(Aminophenyl)
anilines

(6.46)

6.8 Aromatic amines

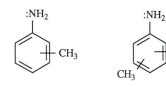

Toluidines   Xylidines

Phenylenediamines

Arylamines are liquids or low melting point solids with limited water solubility. They are toxic and some act as carcinogens. *C*-Alkylanilines have trivial names, thus *C*-methylanilines are known as toluidines and their dimethyl analogues as xylidines.

6.8.1 Synthesis

Arylamines are almost invariably synthesized by the reduction of the corresponding nitro compounds (see p. 49) (eqn 6.47a, b), but in common with the aliphatic series, benzamides can be degraded to anilines under Hofmann conditions — treatment with bromine and potassium hydroxide (eqn 6.48).

Selective reduction of dinitrobenzenes is possible with sodium sulphide or with tin(II) chloride (Fig 6.18).

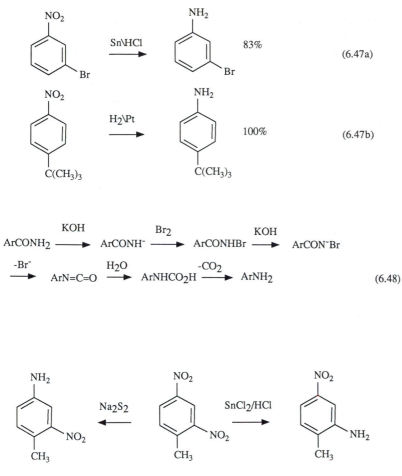

(6.47a)

(6.47b)

$$ArCONH_2 \xrightarrow{\text{KOH} \atop -Br^-} ArCONH^- \xrightarrow{\text{Br}_2} ArCONHBr \xrightarrow{\text{KOH}} ArCON^-Br$$

$$\xrightarrow{-Br^-} ArN=C=O \xrightarrow{\text{H}_2\text{O}} ArNHCO_2H \xrightarrow{-CO_2} ArNH_2 \quad (6.48)$$

Fig. 6.18 Reduction of dinitrotoluene

6.8.2 Basicity

Aniline (pK_b 9.4) is much less basic than methylamine (pK_a 3.4) because of (a) the inductive effect of the phenyl group (sp^2- rather than sp^3-hybridized carbon atoms) and, more importantly, (b) delocalization of the lone pair electrons on the nitrogen atom with the π system (Fig. 6.19). Delocalization provides stability to aniline which is not available to the anilinium cation (eqn 6.49), and simultaneously renders the lone pair electrons less available for protonation. Despite this, aniline is rather more basic than might be predicted. Mutual overlap between the lone pair electrons and the π system would be maximized if the amino group were trigonal with the lone pair occupying a p orbital. This is not the most stable arrangement for an amine, and ammonia, for example, has a pyramidal structure, i.e. the orbitals have a high degree of sp^3 character. In aniline there is a compromise and the lone pair orbital has more p-character than it has in ammonia. Even so the H–N–H plane bisects the plane of the benzene ring at an angle of c. 40 °C and effective delocalization of the lone pair electrons is reduced.

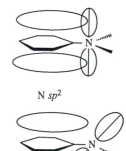

N sp^2

N sp^2-sp^3

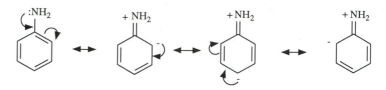

Fig. 6.19 Stabilization of aniline (aminobenzene) by delocalization of the lone pair electrons on the nitrogen atom

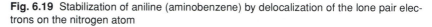

$$\text{(6.49)}$$

Anilinium cation

As in the case of acids (see p. 44), *ortho* substituents exert steric effects and this makes a prediction of basicity difficult. However, the inductive influence of electron withdrawing groups is greatest at the *ortho* positions, although it is still apparent in some *m*-substituted anilines. Unsaturated electron-withdrawing groups at the *o* and *p* positions are conjugated with the lone pair electrons and this resonance effect further reduces their availability for protonation. The basicity of some prominent anilines is recorded in Table 6.2.

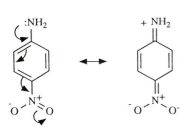

Table 6.2 The basicity of some substituted anilines

Aniline X-C₆H₄-NH₂			
		pK_b	
Substituent X	*o*-	*m*-	*p*-
CH_3	9.6	9.3	8.9
OCH_3	9.5	9.8	8.7
Cl	11.4	10.7	10.2
CN	13.0	11.2	12.3
NO_2	–	11.5	13.0

6.8.3 Anilinium salts

Anilinium chloride and similar salts are readily hydrolysed and their aqueous solutions are acidic to litmus. As a result anilinium chloride is sometimes misleadingly called aniline hydrochloride.

Anilinium hydrogensulphate when heated affords sulphanilic acid (eqn 6.50), which exists as the zwitterion. *N*-Alkylanilinium salts rearrange on heating to give *o*- and *p*-alkylanilines (eqn 6.51).

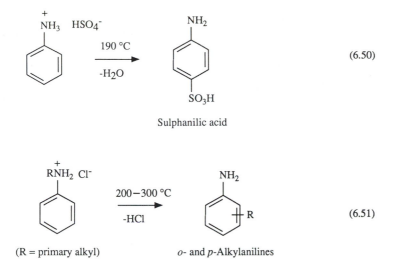

(6.50)

Sulphanilic acid

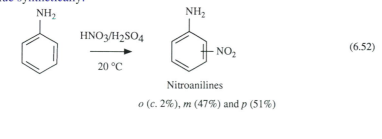

(6.51)

(R = primary alkyl)   *o*- and *p*-Alkylanilines

6.8.4 Reactions with electrophiles

Anilines react readily with electrophiles either at the N atom or in the aromatic nucleus. In general, reactions at the nitrogen atom are favoured because no loss of aromaticity is entailed. Protonation of anilines is an example, and this complicates reactions in acidic media such as nitration and sulphonation (eqn 6.52). Here the substrate for further reaction is principally the anilinium cation and, although the unprotonated amino group is a very strong *o/p* activator, reactions with the anilinium cation proceed relatively slowly to give *m*-oriented products (Fig. 6.20). Mixtures form because, even in strong acid, an equilbirium exists between the anilinium cation and the free base; there is then a competition between a slow reaction with the abundant anilinium cation and a fast reaction between the less available, but replenishable, amount of the free base (see p. 22). Normally these reactions are of little value synthetically.

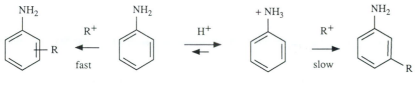

(6.52)

Nitroanilines

o (*c*. 2%), *m* (47%) and *p* (51%)

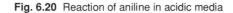

o and *p*

Fig. 6.20 Reaction of aniline in acidic media

With halogens substitution occurs extremely easily and a carrier is not required, thus aniline forms 2,4,6-trichloro and 2,4,6-tribromo derivatives simply by reaction with the appropriate halogen in water (see p. 64). In these cases hypohalous acids XOH (X = Cl or Br) are formed, which are more reactive electrophiles than the halogens themselves (eqn 6.53). It is possible that an *N*-haloaniline is formed first, and rearranges to give a *C*-halogenaniline. Such intermediates are implicated in the chlorination of acetanilides (see below).

$$X_2 + H_2O \quad \rightleftharpoons \quad X^- + X\overset{+}{O}H_2 \quad \rightleftharpoons \quad HX + XOH \qquad (6.53)$$

Anilines are also very prone to oxidation and a sample of freshly distilled aniline rapidly darkens on exposure to air. Some chemical oxidants yield a complex product called aniline black. Potassium dichromate in dilute sulphuric acid gives 1,4-benzoquinone (Fig. 6.21).

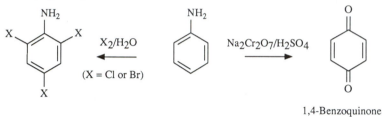

Fig. 6.21 Some reactions of aniline

6.8.5 Anilides

Friedel–Crafts reactions cannot be carried out on anilines because the Lewis acid reacts preferentially with the nitrogen lone pair electrons. However, acid chlorides and anhydrides in the presence of a base (sodium acetate, sodium hydroxide, or a tertiary amine such as pyridine) give the corresponding secondary amides (anilides) (see p. 44).

Anilides are useful substrates for electrophilic substitution because internal resonance within the amide group reduces, but does not totally overcome, the resonance interaction of the nitrogen lone pair with the π system of the ring. As a result acetanilide can be nitrated to afford *o*- and *p*- nitroacetanilides (eqn 6.54). Hydrolysis of the amide group is effected by treatment with dilute sulphuric acid.

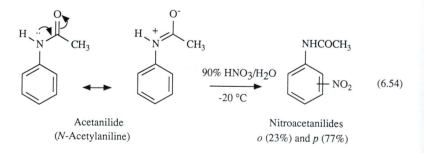

Acetanilide forms mainly *p*-halogeno derivatives on halogenation and it is interesting that when acetanilide is treated with sodium hypochlorite *N*-chloroacetanilide is formed. On reaction with hydrochloric acid this product generates chlorine and acetanilide which in turn react to give *p*-chloroacetanilide, plus a little of the *o*-isomer (eqn 6.55). The rearrangement, known as the Orton reaction, probably involves an ionic mechanism, but irradiation of *N*-chloroacetanilide with ultraviolet light or treatment with radical initiators also produces *p*-chloroacetanilide.

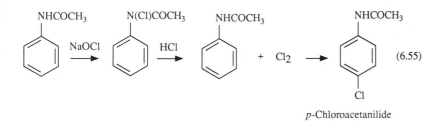

p-Chloroacetanilide

6.8.6 N-Alkylanilines

N-Methylaniline (pK_b 9.9) and *N*,*N*-dimethylamine (pK_b 9.4) are obtained when aniline is alkylated by treatment with methyl halides or dimethyl sulphate. A more useful synthesis of *N*-methylaniline involves the catalytic hydrogenation of the imine formed between aniline and formaldehyde. The two steps of this procedure, can be carried out in the same reaction vessel. Reductive alkylation of anilines is a general reaction and many aldehydes, RCHO, can be used. Instead of hydrogenation, reduction of the intermediate, ArN=CHR, can be achieved using sodium triacetoxyborohydride as the reagent. Another approach to alkylarylamines is the reduction of amides by lithium aluminium hydride. This last method can also be applied to *N*-alkylamides giving in this case *N*,*N*-dialkylarylamines (Fig. 6.22).

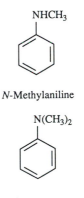

N-Methylaniline

N,*N*-Dimethylaniline

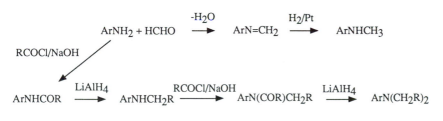

Fig. 6.22 Synthesis of *N*-alkylanilines

6.8.7 Diazonium compounds

N,*N*-Dimethylaniline reacts with sodium nitrite and hydrochloric acid to form *p*-nitroso-*N*,*N*-dimethylaniline (eqn 6.56). The electrophile in this reaction can be thought of as $^+$NO, formed by protonation and then dehydration of nitrous acid. However, this is not certain and nitrous acid is known to exist in equilibrium in aqueous solution with dinitrogen trioxide and also with nitric oxide. Nitrosyl chloride is also present and any one of these species could be involved.

$HNO_2 + H^+ \rightleftharpoons H_2N^+O_2 \rightleftharpoons NO^+ + H_2O$
$2HNO_2 \rightleftharpoons H_2O + N_2O_3$
$3HNO_2 \rightleftharpoons 2NO + HNO_3 + H_2O$
$HNO_2 + HCl \rightleftharpoons NOCl + H_2O$

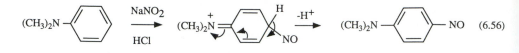

For secondary arylalkylamines such as *N*-methylaniline, the lone pair of electrons on nitrogen can be utilized without disrupting the π system of the aromatic ring, and nitrosation takes place to yield *N*-nitrosoanilinium salts. On deprotonation these afford *N*-nitrosamines, but if the salts are heated in the presence of hydrochloric acid they undergo the Fischer–Hepp rearrangement to yield the appropriate *p*-nitrosamine (Fig. 6.23). The exclusive formation of the *p*-isomer is puzzling! This reaction was once thought to be intermolecular, but this is in doubt since if a nucleophile such as urea is added in large excess the reaction is not inhibited. If $^+$NO, or its equivalent, were formed it would be expected to be effectively scavenged by the abundant nucleophile.

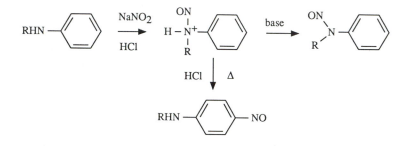

Fig. 6.23 Nitrosation of secondary arylalkylamines

N-Nitrosation is common to both secondary alkylamines and alkylarylamines. It also occurs with primary amines, but in this case initial *N*-nitrosation is followed by dehydration and the production of diazonium salts (eqn 6.57). Aliphatic diazonium salts rapidly decompose, a reaction facilitated by the loss of the stable molecule nitrogen which is amongst the best of all leaving groups. Aromatic diazonium salts are more stable due to delocalization of the charge on nitrogen with the π system (Fig. 6.24), and aqueous solutions of diazonium salts can be prepared and used at 0–5 °C. The solids tend to be explosive, but additional stability is noted when a non-polarizable counterion is present. Diazonium tetrafluoroborates and hexafluorophosphates can sometimes be isolated and handled with a degree of safety.

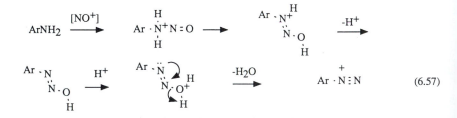

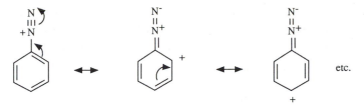

Fig. 6.24 Stabilization of aromatic diazonium salts

Benzene diazonium salts are very useful in the synthesis of a wide range of aryl derivatives. Whereas the S_N1 nucleophilic displacement of aryl halides is unknown, the decomposition of benzene diazonium salts is a first order process and the benzene carbocation is produced. Once formed the cation is very reactive and combines readily with available nucleophiles to afford substituted benzenes. The driving force for this reaction is the elimination of the very stable nitrogen molecule (see above), and evidence for the participation of the benzene cation is obtained from the observation that when $Ar^{15}N^+ \equiv N$ is the reactive species, and the opportunity for reactions with nucleophiles are minimized, recovered starting material consists of $Ar^{15}N^+ \equiv N$ and $ArN^+ \equiv {}^{15}N$.

It is possible that 'aryl cations' are not simple structures for there is evidence that they also contain unpaired electrons, i.e. they show radical characteristics.

Benzene diazonium salts react with water to give phenols, and with aqueous solutions containing halide ions to yield halobenzenes (Fig. 6.25). If two ions of similar nucleophilicity are present then mixtures form: thus when diazonium chlorides are hydrolysed mixtures of phenols and chlorobenzenes are produced. For this reason if phenols are required it is usual to carry out the diazotization of the amine in sulphuric acid rather than in hydrochloric acid. Potassium ethyl xanthate with aryldiazonium salts gives arylethyl xanthates, which on hydrolysis produce thiophenols (Fig. 6.25) (this is a safer method than using metal polysulphides).

Alkali metal azides yield azidobenzenes, and the reduction of benzene diazonium salts to the parent hydrocarbons is conveniently achieved by the use of hypophosphorus acid catalysed by copper, although sodium stannite and alkaline formaldehyde are also recommended.

The reduction may involve radicals, see below.

Certain cuprous salts — chloride, bromide, and cyanide — catalyse the decomposition of benzene diazonium salts and give the corresponding chloro-, bromo- or cyano-benzenes, respectively. Collectively these reactions are known as Sandmeyer reactions, and they involve a homogeneous rather than a heterogeneous fragmentation of the aryl–nitrogen bond. The cuprous ion acts to reduce the diazonium salt

Cuprous iodide is very insoluble and cannot be used in a Sandmeyer reaction.

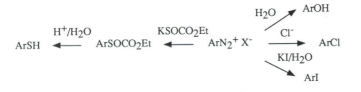

Fig. 6.25 Reactions of benzene diazonium salts

and an aryl radical is formed which donates an electron to cupric ion reducing it back to the cuprous state. The cuprous salt is thus a true catalyst (eqn 6.58a, b).

$$ArN_2^+X^- + CuX \longrightarrow Ar\cdot + N_2 + CuX_2 \tag{6.58a}$$

$$Ar\cdot + CuX_2 \longrightarrow Ar\,X + CuX \tag{6.58b}$$

Fluorobenzenes cannot be made in this way but they are available through the Bälz–Schiemann reaction in which an aryl diazonium tetrafluoroborate or pentafluorophosphate, is cautiously heated. Nitrobenzenes can be prepared by treating benzene diazonium tetrafluoroborates with aqueous sodium nitrite in the presence of cuprous ions (eqn 6.59).

$$ArNO_2 \xleftarrow{\;NaNO_2/Cu^+\;} ArN_2^+BF_4^- \xrightarrow{\;\Delta\;} ArF + N_2 + BF_3 \tag{6.59}$$

Benzene diazonium salts are also weak electrophiles and react (couple) with electron rich arenes (mainly phenols and amines) at the *para* position to give azo compounds. The coupling reaction with phenols must be carried out in alkaline media, so that in fact phenate anions are the substrates. In the case of primary amines, diazonium salts react faster with the lone pair of electrons of the amino group than with the ring carbon atoms, so that the initial products are *N*-diazoaminobenzenes (triazines). Such compounds are thermodynamically less stable than the aminoazobenzene isomers and when they are heated with anilinium salts they rearrange. This is usually regarded as a reversal of the diazoaminobenzene formation, but it could be more complex (Fig. 6.26).

Azo compounds, because of their highly conjugated structures are brightly coloured and are used as dyestuffs. In the past, some were used as food colouring agents — butter yellow and BFK (brown for kippers) are two examples — but the possibility that such compounds might be carcinogenic has discouraged this practice.

Benzene diazonium salts, when prepared in the minimum of water and the solutions then basified with sodium hydroxide, give aryldiazohydroxides. These can be extracted into an organic solvent whereupon they decompose to aryl radicals, which dimerize and lose hydrogen to give biaryls (eqn 6.60a, b). This process, known as the Gomberg–Backmann reaction, is similar to the Ullmann reaction (p. 36) and can sometimes be of synthetic value. However, there are many alternative pathways by which the radicals can react and yields are not good.

BFK   (A r = *p*-HO$_3$SC$_6$H$_4$-)

Butter yellow

$$ArN_2^+\,Cl^- + HO^- \xrightarrow{\;-Cl^-\;} ArN=NOH \longrightarrow Ar\cdot + N_2 + HO\cdot \tag{6.60a}$$

$$2Ar\cdot \longrightarrow Ar\text{-}Ar \tag{6.60b}$$

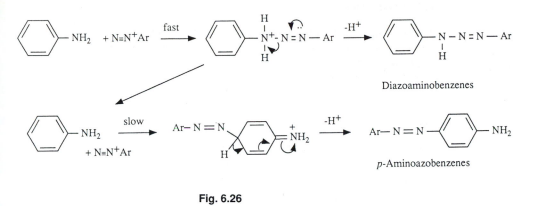

Diazoaminobenzenes

p-Aminoazobenzenes

Fig. 6.26

6.9 Phenols

6.9.1 Nomenclature and occurrence

Phenols are compounds in which one or more hydroxyl groups are directly bonded to the aromatic nucleus. The parent compound phenol and the methylphenols (cresols) are found in coal tar, and other monohydric phenols and polyhydric phenols are common constituents of plant extracts. They frequently co-occur with their methyl ethers. Many have trivial names and some examples are cited in Fig. 6.27.

The names phenols and phenates (the anions formed on treatment of phenols with alkali, see below) serve to emphasize their relationships to enols and enolates respectively.

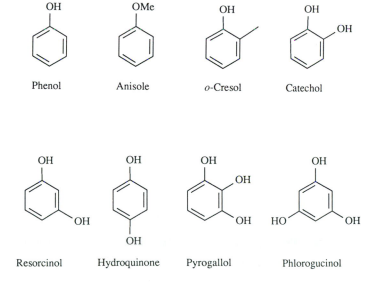

| Phenol | Anisole | *o*-Cresol | Catechol |

| Resorcinol | Hydroquinone | Pyrogallol | Phlorogucinol |

Fig. 6.27 Some examples of phenols and analogues

6.9.2 Synthesis

Cumene is synthesized by reacting benzene with propene in the presence of a strong acid, see p. 15.

Phenol may be synthesized industrially by the oxidation of isopropylbenzene (cumene). Initially a hydroperoxide is produced, which undergoes acid-catalysed fragmentation and rearrangement to an oxonium ion. Hydrolysis of this species then affords phenol and acetone. (eqn 6.61).

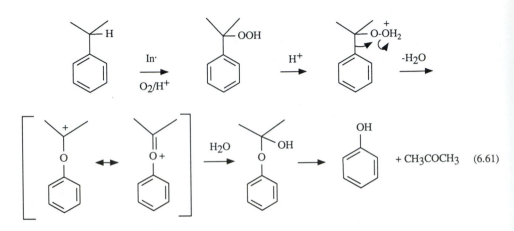

Older methods include heating chlorobenzene with sodium hydroxide/super-heated water at 250 °C and 13.8–20.7 x 10^6 Pa, or fusing a benzenesulphonic acid with either sodium or potassium hydroxide (eqn 6.62) (see p. 29). The conditions of these reactions are drastic and they are only successful if any other groups which are present are able to survive them. Phenols can also be made by the hydrolysis of diazonium salts (eqn 6.63) (see p. 59).

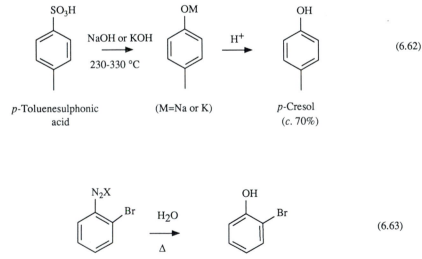

6.9.3 Acidity and structure

One of the lone pair electrons of the hydroxyl group of phenol overlaps with the π system of the ring and this extra delocalization contributes about 8 kJ to the overall resonance energy of the molecule (Fig. 6.28).

Phenols give rise to coloured complexes when treated with ferric chloride. These colours are often diagnostic for a particular phenol, and there is frequently a variation in the colour depending upon whether the test is carried out with water or alcohol as the solvent. Phenol gives an acid reaction to litmus and dissolves in dilute sodium hydroxide solution. This shows up a marked difference in chemistry between phenols and alcohols, which are neutral compounds (phenol pK_a 10; cyclohexanol pK_a 18). If electron withdrawing substituents, such as nitro groups, are present at the *ortho* and *para* positions the acidity of the phenol is increased (4-nitrophenol pK_a 7.2). The acidity of phenol can be attributed to the stability of the phenate anion in which the formal negative charge on the oxygen atom is delocalized throughout the ring (Fig. 6.29). *Ortho-* or *para-* nitro groups allow further extensions of this effect.

The enhanced acidity of *o-* and *p-*nitrophenols is demonstrated by their solubility in aqueous carbonate solution. 2,4,6-Trinitrophenol (picric acid) (pK_a 0.8) is as acidic as many mineral acids and can be used to etch metals. *m-*Nitrophenol is insoluble in sodium carbonate but dissolves in sodium hydroxide: here, of course, the nitro group and the lone pair electrons of the hydroxyl substituent are not conjugated.

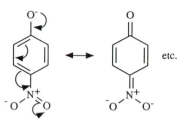

Intramolecular hydrogen bonding within *ortho*-substituted phenols reduces their vapour pressure relative to that of the *m* and *p* isomers. For example, *o*-nitrophenol can be steam-distilled, whereas its isomers cannot. Hydrogen bonding in the *ortho* compounds also leads to an increase in acidity by weakening the hydroxyl bond.

etc.

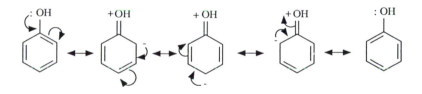

Fig. 6.28 Delocalization in phenol

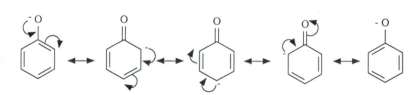

Fig. 6.29 Delocalization in the phenate anion

6.9.4 Chemical reactions

Although rapid electrophilic monobromination is expected, subsequent brominations should be more difficult. The ease with which the tribromo derivative forms suggests that a complex reaction mechanism operates. It will be recalled that aniline too forms a tribromo derivative under similar conditions (see p.56).

The resonance effect of the lone pair electrons of the hydroxyl group increases electron density in the benzene nucleus promoting electrophilic attack and similarly lowering the energy contents of the relevant *ortho* and *para* sigma intermediates. This is reflected in the ease of electrophilic–substitution reactions. Phenol reacts with bromine in carbon tetrachloride to give *p*-bromophenol and no Lewis acid is required (eqn 6.64). With bromine water 2,4,6-tribromophenol is formed (eqn 6.65).

Nitration occurs with 20% aqueous nitric acid, below 30 °C, to give mainly *o*- and *p*-nitrophenols (eqn 6.66). The concentration of $^+NO_2$ in this medium is too low to explain the ease of reaction, and it has been shown that the active reagent is

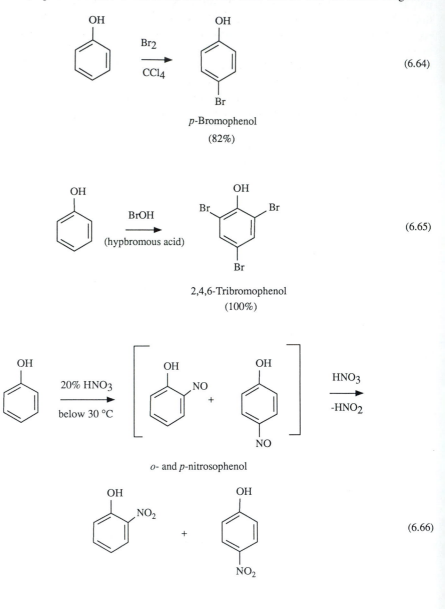

(6.64)

p-Bromophenol
(82%)

2,4,6-Tribromophenol
(100%)

o- and *p*-nitrosophenol

(6.65)

(6.66)

actually nitrous acid. This nitrosates phenol through the agency of the nitrosonium ion ^+NO. The *o*- and *p*-nitrosophenols which are produced are then rapidly oxidized by nitric acid to the corresponding nitrophenols. In this last reaction nitric acid is reduced to nitrous acid, which increases in concentration thus accelerating the first reaction.

Nitrous acid alone reacts with phenol at 0 °C to give *o*- and *p*-nitrosophenols. The *para* derivative is tautomeric with the mono-oxime of 1,4-benzoquinone (eqn 6.67).

When phenol is deprotonated the phenate anion is produced in which electrophilic reactivity is enhanced further, as now the resonance effect involves a full negative charge. Phenates react readily with a range of relatively weak electrophiles. For example, sodium phenate combines with formaldehyde to give *o*- and *p*-benzyl alcohols (Lederer–Manasse reaction). If the reaction is carried out in an acidic medium these products react further to produce a complex phenol–formaldehyde resin (Bakelite) (eqn 6.68).

When heated at 130 °C under pressure carbon dioxide reacts with sodium phenate to give mainly *o*-hydroxybenzoic acid (salicylic acid). The reaction depends on the nature of the cation: if potassium phenate is used at a temperature of 180 °C the *p*-isomer predominates. The reaction is known as the Kolbe process (eqn 6.69).

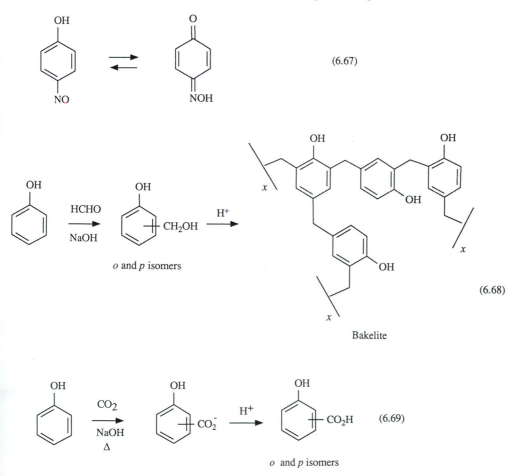

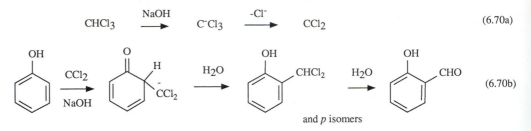

$$\text{CHCl}_3 \xrightarrow{\text{NaOH}} \text{C}^-\text{Cl}_3 \xrightarrow{-\text{Cl}^-} \text{CCl}_2 \qquad (6.70a)$$

and *p* isomers (6.70b)

Another reaction with a long history is the Riemer–Tiemann reaction in which a phenol is treated with alkali and chloroform to give the corresponding *o*- and *p*-hydroxybenzaldehydes. The mechanism of this reaction puzzled chemists until it was shown that the electron deficient molecule dichlorocarbene is formed by the base promoted dehydrochlorination of chloroform (eqns 6.70a,b).

In the Lederer–Manasse, the Kolbe and the Riemer–Tiemann reactions the *ortho* product predominates over its *para* isomer. This is due to chelation control, which is only possible within the *ortho* dipolar reaction intermediates, and has the effect of lowering energy by delocalizing the charges. A change from sodium hydroxide to potassium hydroxide in the Kolbe reaction means that the less polarizable potassium ion replaces the sodium cation in the intermediate. Chelation is less effective and the *para* isomer is preferred.

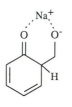

Lederer - Manasse

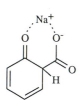

Kolbe

Riemer - Tiemann

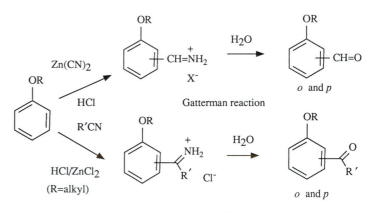

Gatterman reaction

Hoesch (Houben - Hoesch) reaction

Fig. 6.30 The Gatterman and Hoesch reactions

Phenols react under Gatterman conditions to give aldehydes. The similar Hoesch reaction fails with phenols, but succeeds with phenolic ethers affording ketones (Fig. 6.30). In both cases the reactions work best when the substrates bear electron releasing substituents.

Phenols are *O*-alkylated (affording ethers) by reaction with dialkyl sulphates or alkyl halides, and *O*-acylated (to give the corresponding esters) by treatment with acyl halides or anhydrides (Fig. 6.31). Both types of reaction require the presence of a base, commonly sodium hydroxide or sodium acetate, and are further examples

In the case of the phenols the Hoesch reagent attacks at the oxygen atom to give imino ethers.

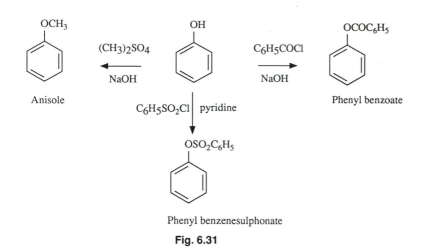

Fig. 6.31

of the Schotten–Baumann procedure (see p. 44). The reactions can be made homogenous by using a tertiary amine, such as pyridine, as the base. *O*-Sulphonylation is carried out in similar fashion.

O-Acylations can also be promoted by a trace of acid. *O*-Acetylsalicylic acid, better known as aspirin, is formed by heating salicylic acid with acetic anhydride and a catalytic amount of suphuric acid. (eqn 6.71).

Phenols form complexes with Lewis acids, such as aluminium trichloride, through reaction of the electron deficient Lewis acid with the lone pair electrons on the oxygen atom. Even when Friedel–Crafts acylation reactions are carried out using alkoxybenzenes similar complexation is possible and the products may be *O*-dealkylated under the reaction conditions.

Phenolic esters rearrange in the presence of aluminium trichloride (Fries rearrangement), which initially reacts with the lone pair electrons of the oxygen atom to generate an acylium species (analogous to that formed in the Friedel–Crafts reaction between an acyl choride and aluminium trichloride). This then attacks the ring (eqn 6.72). The reaction may be conducted under thermodynamic or kinetic control, hence at high temperature the *ortho* product is favoured as a result of stabilization through chelation with the aluminium.

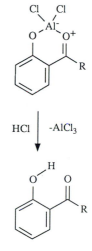

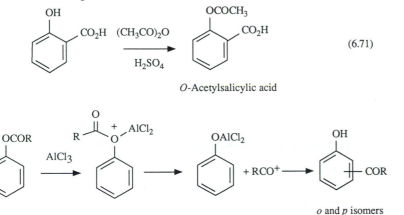

(6.71)

O-Acetylsalicylic acid

(6.72)

o and *p* isomers

The Fries rearrangement is intermolecular, but when allyl phenyl ethers are heated *C*-allylphenols are formed. The reaction known as the Claisen rearrangement follows an intramolecular course and is an example of a pericyclic [3,3]sigmatrophic rearrangement (eqn 6.73).

The Claisen rearrangement can be more complex than that shown in eqn 6.73. For example, if the *o* positions of the starting material are occupied the initially formed dienone may undergo a further [3,3]sigmatropic shift in which the allyl group migrates to the 4-position. Aromatization then affords a 4-allylphenol. Should the *o* and *p* sites be fully subsituted then the allyl group 'cartwheels' around the ring and ends up back on the oxygen atom.

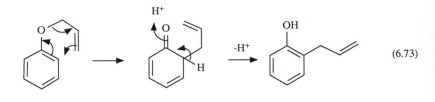

(6.73)

6.9.5 Oxidation

Phenols are prone to oxidation, frequently giving rise to complex mixtures. Sometimes a degree of control can be exerted and specific products can be obtained. For example, the oxidation of catechol and hydroquinone gives 1,2- and 1,4-benzoquinones, respectively (eqn 6.74a, b)

Phenate ions are particularly susceptible to oxidation and the initial products are the corresponding phenoxy radicals. These enter into self-coupling reactions to form hydroxylated biphenyls, which are further oxidized to quinones (*para–para* coupling of phenoxy radicals is most frequently observed) (eqn 6.75).

Unfortunately these are not the only products, and such reactions are notorious for the production of intractable tars. Thus although there have been many attempts to effect intramolecular cyclizations through the oxidative coupling of phenols in alkaline media, the yields of products are usually disappointing. Better results come from the oxidation of *O*-alkylphenols, where it is likely that the reactive species are radical cations. These can be generated electrochemically, or with the one-electron oxidants, as, for example, in the oxidative cyclization of metacyclones to bis-dienones by reactions with ferric chloride (eqn 6.76).

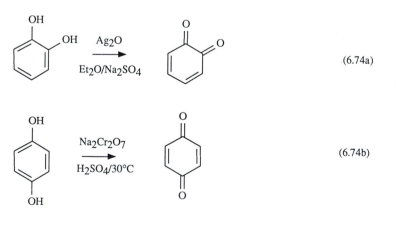

(6.74a)

(6.74b)

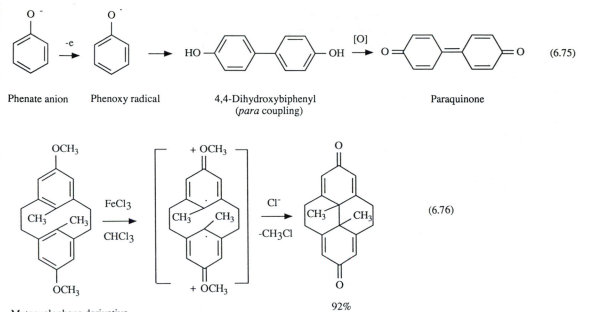

Phenate anion Phenoxy radical 4,4-Dihydroxybiphenyl
(*para* coupling) Paraquinone (6.75)

Metacyclophane derivative 92% (6.76)

7 Polycyclic arenes

7.1 Introduction

Another allotrope of carbon, buckmaster fullerene (C_{60}), consists of spherically-fused five and six rings of carbon atoms, assembled like the panels of a soccer ball (hence the popular name 'bucky ball' for this compound).

The fusion of benzene rings leads to an array of polycyclic arenes, of which naphthalene, anthracene, and phenanthrene, are the best known. Other compounds in the series include chrysene, pyrene, tetracene, and coronene (Fig. 7.1).

Many polycyclic aromatic hydrocarbons are found in coal tar, and are formed when plant materials are partially combusted. Some are procarcinogens, being oxidized within the body to the true cancer promoting compounds. Very large polycyclic systems are known, and graphite consists of planes derived from an infinite linear fusion of benzene nuclei, each plane being separated from its neighbour by a distance of 3.4 Å.

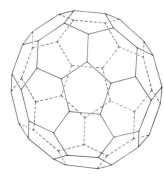

Naphthalene   Anthracene   Phenanthrene

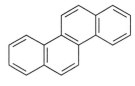

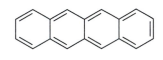

Chrysene   Pyrene   Tetracene

Coronene

Fig. 7.1 Some polycyclic arenes

7.2 Naphthalene

7.2.1 Structure and nomenclature of naphthalene and its derivatives

The naphthalene ring is numbered as shown, but the C-1 and C-2 positions are commonly referred to as α and β respectively (Fig. 7.2). The α sites are also known as *peri* positions.

Naphthalene is much more easily hydrogenated than benzene (see below), but the reduction only proceeds as far as 1,2,3,4-tetrahydronaphthalene (tetralin). Tetralin resembles 1,2-dimethylbenzene, and further hydrogenation requires high pressure and elevated temperatures. This result is fully in accord with the aromatic character of naphthalene, but the empirical (or stabilization) energy (see p. 3) of naphthalene (241 kJ mol^{-1}) is not twice that of benzene (172 kJ mol^{-1}): a consideration of the main contributors to the valence bond description of naphthalene show that it is not possible for both rings to be benzenoid at the same time (Fig. 7.3). Carbon–carbon bond lengths also reflect this and, in particular, the 1,2-bond is shorter (137 pm) (more double bond character) than the 2,3-bond (140 pm).

Note that there are four equivalent α and four equivalent β sites.

C–C bond lengths in benzene are all 140 pm.

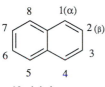

Naphthalene

1,4-Dihydronaphthalene

1,2,3,4-Tetrahydronaphthalene
(tetralin)

Decahydronaphthalene
(decalin)

Fig. 7.2 Naphthalene and its derivatives

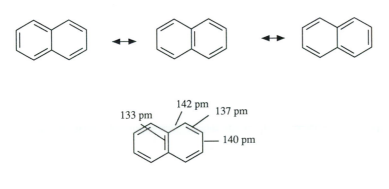

Fig. 7.3

7.2.2 Synthesis

A classical synthesis of naphthalene starting from benzene employs a Haworth succinoylation reaction (see p. 15) as the first step. This Friedel–Crafts type procedure affords 4-phenyl-4-oxobutanoic acid, which is reduced by a modified Wolff–Kishner reduction (utilizing hydrazine, glycol, and potassium hydroxide as reagents) to 4-phenylbutanoic acid. The acid is then cyclized to 1-oxo-1,2,3,4-tetra-hydronaphthalene (α-tetralone) by treatment with polyphosphoric acid. Reduction of the ketone group of α-tetralone, as for 4-phenyl-4-oxobutanoic acid, gives tetralin; finally, oxidative aromatization (dehydrogenation) to naphthalene is effected by heating tetralin with a palladium catalyst (Fig. 7.4). The synthesis can be applied to other naphthalene derivatives, providing the substituents in the benzene ring are not deactivating. In addition, α-tetralone allows a further extension of the approach since reactions of it with Grignard reagents allows access to 1-substituted naphthalenes.

Another important route to naphthalenes employs Diels–Alder cycloadditions between 1,4-benzoquinone and 1,3-dienes. The adducts which form can be enolized to dihydroxydihydronaphthalenes, and then oxidized to 1,4-napthoquinones (eqn 7.1).

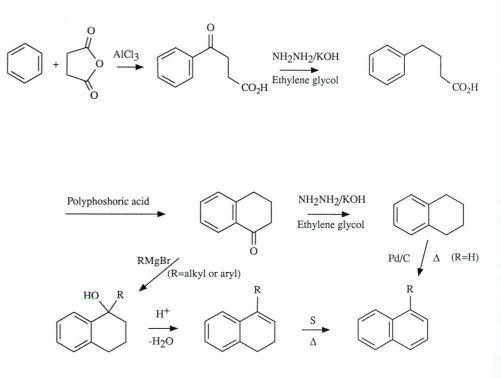

Fig. 7.4 Synthesis of naphthalene

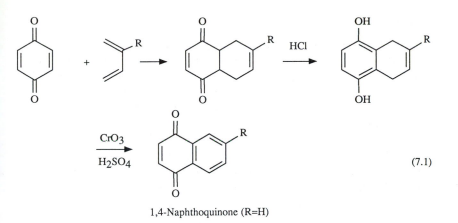

(7.1)

1,4-Naphthoquinone (R=H)

7.2.3 Electrophilic substitution reactions

7.2.3.1 Monosubstitution

Under conditions favouring kinetic control, naphthalene reacts with electrophiles (Y^+) faster at the α position. The selection of the α site follows from an analysis of the valence bond contributors to the appropriate sigma complexes: thus the intermediate corresponding to attack at C-1 has two contributions in which one benzenoid ring remains intact, whereas that for C-2 attack has only one (Fig. 7.5). These contributions have lower energies than the others where both benzene π systems are disrupted. The fact that the wave equation describing the α intermediate has two such low energy factors accounts for the lower energy of this sigma complex.

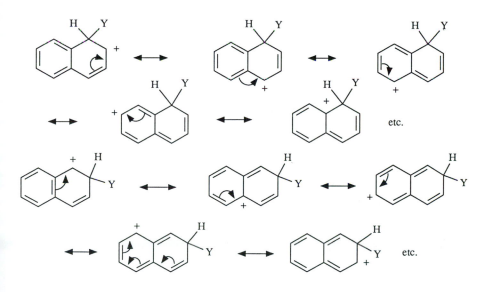

Fig. 7.5

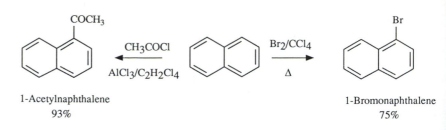

1-Acetylnaphthalene
93%

1-Bromonaphthalene
75%

Fig. 7.6 Some reactions of naphthalene with electrophiles

This reflects the reduction in overall resonance stabilization energy when two benzenoid units are fused together, plus the fact that in the sigma complexes for monosubstitution, at least, the unattacked nucleus retains its aromatic character.

Electrophiles tend to react with naphthalene more rapidly than with benzene. For example, bromine in boiling carbon tetrachloride affords *c.* 75% 1-bromonaphthalene and no carrier is necessary. Similarly, Friedel–Crafts acylation with acetyl chloride and aluminium trichloride, in tetrachloroethane as solvent and under mild conditions, affords mainly 1-acetylnaphthalene (Fig. 7.6). If the Friedel–Crafts reaction is repeated in nitrobenzene, rather than in tetrachloroethane as solvent, a 90% yield of 2-acetylnaphthalene is obtained (see below).

Nitration with nitric acid in acetic anhydride at 60 °C gives 1-nitronaphthalene and 2-nitronaphthalene in the ratio 10:1, plus smaller amounts of 1,8- and 1,5-dinitronaphthalenes (eqn 7.2). The dinitronaphthalenes are the products of further nitration of 1-nitronaphthalene. It is noteworthy that the presence of the deactivating nitro group in one ring directs further electrophilic attack into the unsubstituted benzene nucleus (see below).

Although attack at C-1 is faster than at C-2, a number of examples exist where the kinetically favoured α-subsitituted naphthalene reverts to a thermodynamically more stable β isomer if the temperature is raised and sufficient time is allowed for equilibration to occur. Such illustrations of thermodynamic versus kinetic control are important since they provide access to other β-functionalized naphthalenes.

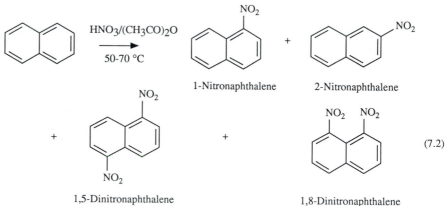

1-Nitronaphthalene

2-Nitronaphthalene

1,5-Dinitronaphthalene

1,8-Dinitronaphthalene

(7.2)

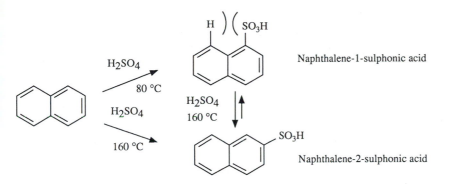

Fig. 7.7 Sulphonation of naphthalene

The isomerization is commonly noted when the substituent group is large. This is because the fused nature of naphthalene dictates that there is steric hindrance between the hydrogen atom at C-8 and a group at C-1. At C-2 the substituent is immediately flanked by two protons, as in a monosubstituted benzene derivative, and the steric problem is less. The size of the incoming group is also important if steric clashes are to be minimized. In the Friedel–Crafts acetylation reactions discussed above, for example, the change to nitrobenzene as solvent, which has a high dielectric constant, is thought to allow more effective solvation of the Lewis acid/reagent complex, thereby increasing its size over that formed in the less polar solvent tetrachloroethane. As a result reaction at C-2 is favoured over attack at C-1.

The sulphonation of naphthalene, provides a good example of thermodynamic versus kinetic control. In sulphuric acid at 80 °C the major product (96%) is naphthalene-1-sulphonic acid. If this compound is heated in sulphuric acid at 160 °C, or if the reaction temperature is raised to about 160 °C, naphthalene-2-sulphonic acid becomes the predominant product (86%) (Fig. 7.7).

The same rules of activation and deactivation apply to substituent groups in both the benzene and the naphthalene series. Electron-donating groups facilitate electrophilic attack, whereas naphthalenes bearing unsaturated electron-withdrawing substituents react more slowly than naphthalene itself. In the latter case the incoming electrophile is directed to the unsubstituted ring, whereas the reverse is true if the substituent activates the molecule to attack.

7.2.3.2 Orientation preferences for monosubstituted naphthalenes

As for monosubstituted benzenes (p. 18), the orientation preferences for the entry of an electrophile into a naphthalene which already bears a single substituent can be assessed by a consideration of the extent of charge delocalization in the sigma complexes. Thus substitution at either C-2 or C-4 is favoured for a naphthalene bearing an electron donating group X at C-1. Should such a group be at C-2 then attack is faster at C-1. This follows because if the electrophile became bonded to C-3, the aromaticity of the other ring would be disturbed (Fig. 7.8).

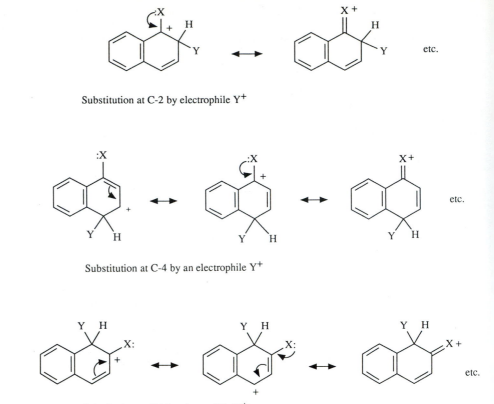

Substitution at C-2 by electrophile Y^+

Substitution at C-4 by an electrophile Y^+

Substitution at C-1 by electrophile Y^+

Substitution at C-3 by an electrophile Y^+

Fig. 7.8 Sigma complexes for electrophilic substitution of monosubstituted naphthalenes

7.2.4 Oxidation and reduction

Aerial oxidation of naphthalene with vanadium pentoxide at 500–600 °C gives phthalic acid and thence phthalic anhydride (see p. 47), whereas treatment with chromium trioxide in acetic acid at 25 °C yields 1,4-naphthoquinone.

Naphthalene undergoes reduction with sodium in ethanol at 78 °C to produce 1,4-dihydronaphthalene; however, at 132 °C, in boiling isopentyl alcohol, tetralin is formed. The reaction involves an electron-addition-protonation sequence — as is usual with dissolving metals. As mentioned above, further reduction of tetralin is difficult, but hydrogenation at high pressure, over a transition metal catalyst (e.g. Rh/C) affords decalin (eqn 7.3).

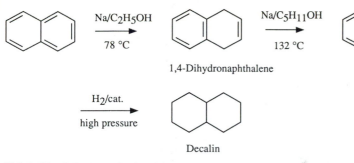

Decalin exists in two form, *cis* (b.p. 195 °C and *trans* (b.p. 185 °C).

cis-Decalin

trans-Decalin

(7.3)

7.2.5 Naphthalene derivatives

Many naphthalene derivatives can be synthesized by routes similar to those used in the benzene series. In general, the substituent groups also react in the same way. Thus halonaphthalenes form lithium and Grignard reagents, which then show comparable chemistry to that of their monocyclic analogues. The halogen atoms are resistant to nucleophilic displacement, but halogenated naphthalenes yield naphthynes under severe conditions (Fig. 7.9).

7.2.5.1 Naphthylamines

Aminonaphthalenes (naphthylamines) are formed from the corresponding nitro compounds by reduction, or from the hydroxynaphthalenes (naphthols) by the Bucherer reaction. In this process, not normally exhibited in the benzene series, a naphthol is heated with ammonia, or a primary amine (RNH$_2$) and sodium hydrogensulphite. The reaction involves an overall addition–elimination mechanism and is only possible because of the reduced aromatic character of the naphthalene system. In the first step protonation at C-4 occurs and then hydrogensulphite anion adds at C-3 to form an enol, which is in tautomeric equilibrium with its keto form. The latter reacts with ammonia, or the amine, to give an imine ⇌ enamine pair, which eliminates sodium hydrogensulphite. The reactions are reversible and thus also provide a route from naphthylamines to naphthols (Fig. 7.10).

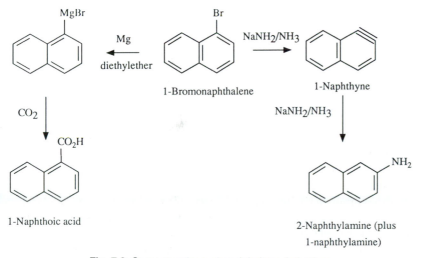

Fig. 7.9 Some reactions of naphthalene derivatives

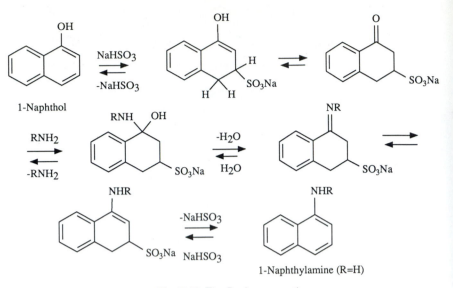

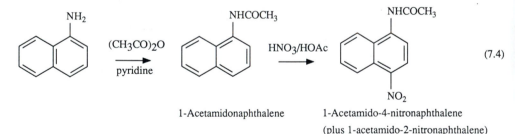

Fig. 7.10 The Bucherer reaction

A similar reaction occurs with 2-naphthol giving 2-naphthylamine (β-naphthy-lamine). This compound was once important in the dyestuffs industry, and its diazonium salt was used to form a wide range of azo dyes. Unfortunately, 2-naph-thylamine is now known to be a carcinogen. Naphthylamines are weak bases which behave in a similar manner to aniline. They react readily with electrophiles and are easily oxidized to naphthoquinones. As with aniline, electrophilic substitution reactions are best carried out using the N-acylated derivatives. For example, 1-ace-tamidonaphthalene can be nitrated with nitric acid in acetic acid to afford a mixture of 2- and 4-nitro-1-acetamidonaphthalenes (eqn 7.4).

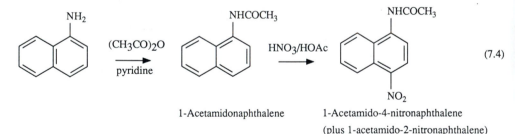

(7.4)

1-Acetamidonaphthalene   1-Acetamido-4-nitronaphthalene
                           (plus 1-acetamido-2-nitronaphthalene)

7.2.5.2 The naphthols

The hydroxy equivalents of the naphthylamines are the naphthols. The parent compounds, 1-(α-) and 2-(β-) naphthols, are usually prepared by fusing the sodium salt of the corresponding sulphonic acid with sodium hydroxide (eqn 7.5). Mechan-istically these reaction *may* occur through the addition of hydroxide ion to the *ipso* position of the sulphonate, followed by the elimination of sulphinate ion. This suggestion is not proven, although the participation of a naphthyne seems to be ruled out because of the regiointegrity of the reaction.

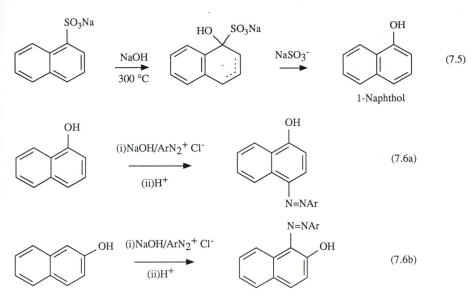

The naphthols are typical phenols and dissolve in alkali, but not in aqueous sodium carbonate solution. They produce colours with ferric chloride, and can be *O*-alkylated by reaction with alkyl halides or with dialkyl sulphates in basic media. Similarly, *O*-acylations can be effected under Schotten–Baumann conditions. Coupling occurs when alkaline solutions of the naphthols are treated with diazonium salts (eqn 7.6a, b), and they undergo other electrophilic reactions easily. For example, naphthols react with nitrous acid to give *C*-nitroso naphthalenes, which are in equilibrium with their oximino tautomers (Fig. 7.11).

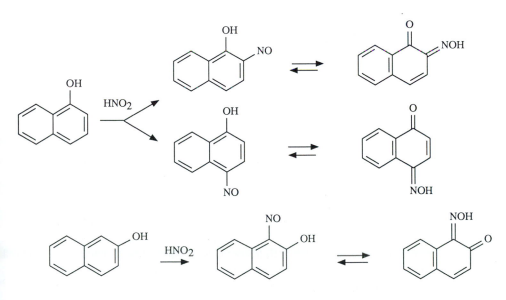

Fig. 7.11 Reaction of naphthols with nitrous acid

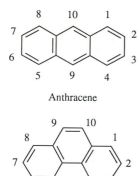

Anthracene

Phenanthrene

7.3 Anthracene and phenanthrene

7.3.1 Anthracene

7.3.1.1 Synthesis

Anthracene can be synthesized in several ways, two of which parallel the constructions of naphthalenes discussed in the previous section. For example, if benzene is reacted with phthalic anhydride and aluminium trichloride *o*-benzoylbenzoic acid is formed, which on treatment with sulphuric acid affords 9,10-anthraquinone (anthraquinone), one of the more important anthracene derivatives. Distillation of anthraquinone with zinc dust yields anthracene (eqn 7.7). A Diels–Alder approach can also be adopted, and if 1,4-naphthoquinone is reacted with 1,3-butadiene a tetrahydroanthraquinone is produced (eqn 7.8). This can be oxidized to anthraquinone by treatment with chromium trioxide in acetic acid.

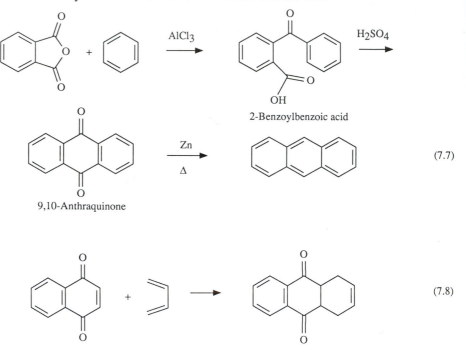

2-Benzoylbenzoic acid

9,10-Anthraquinone

(7.7)

(7.8)

7.3.1.2 Structure

The major contributors to the valence bond description of anthracene are shown in Fig. 7.12. As for naphthalene, these reveal that it is not possible for all the rings to be benzenoid simultaneously, and, moreover, that a four carbon atom fragment of the central ring terminated on either side by C-9 and C-10 has a high degree of 'dienic' character. Indeed, this is manifested in much of the chemistry of anthracene, where additions across the 9,10-positions are commonplace giving 9,10-disubstituted dihydroanthracenes in which the two 'outside' rings are fully benzenoid. It was this special reactivity of anthracene which caused early chemists to emphasize the reactive sites of the molecule and to adopt the unusual numbering system for anthracene which persists to this day.

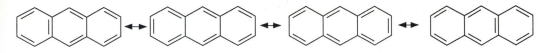

Fig. 7.12 Valence bond descriptions of anthracene

7.3.1.3 Chemical reactions

Anthracene is easily oxidized to 9,10-anthraquinone (anthraquinone) and reduced to 9,10-dihydroanthracene (Fig. 7.13). The molecule acts as a diene and forms adducts in which the dienophile bonds across the 9,10-positions. Benzyne, for example, yields triptycene (see p. 30). This mode of addition is adopted since it allows the flanking rings to each become 6π-electron systems in their own right.

Similarly electrophiles (Y$^+$) react preferentially at C-9 since the carbocationic intermediate thus formed is stabilized by resonance with both of the flanking benzene nuclei. The same intermediate can also readily trap available nucleophiles (X$^-$) at C-10. In this case the overall result is an addition process (Fig. 7.14).

Note that in anthracene all three rings have to share 14 electrons to achieve aromaticity: no one single ring has a 4n+2 π-electron monopoly.

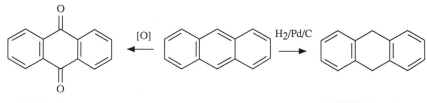

9,10-Anthraquinone

9,10-Dihydroanthracene

Fig. 7.13 Oxidation and reduction of anthracene

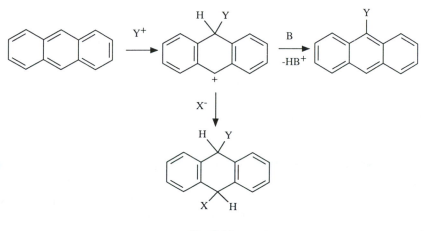

Fig. 7.14

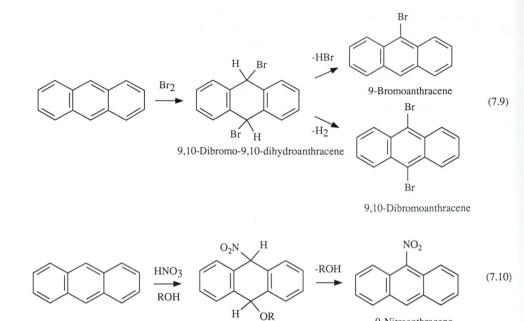

(7.9)

(7.10)

Chlorine and bromine, in carbon disulphide at 0 °C, give dihalides initially, but if the temperature is allowed to rise dehydrohalogenation occurs to yield 9-halogenoanthracenes. Some dehydrogenation also takes place and the product is contaminated with the appropriate 9,10-dihaloanthracene (eqn 7.9). 9-Bromoanthracene is also obtained when anthracene is treated with thallium(III) acetate and bromine (see p. 17).

Concentrated nitric acid oxidizes anthracene to anthraquinone, whereas dilute nitric acid, or nitric acid in acetic acid (ROH = CH$_3$CO$_2$H), affords 10-hydroxy- or 10-acetoxy-9-nitro-9,10-dihydroanthracenes, respectively. These compounds lose water or acetic acid, respectively, on treatment with alkali to yield 9-nitroanthracene (eqn 7.10).

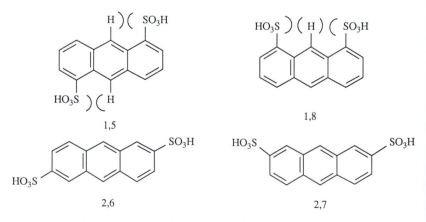

Fig. 7.15 Anthracene disulphonic acids

Large groups at C-9 suffer steric hindrance from both the hydrogen atoms at positions 1 and 8 (*peri* positions). Thus, although attack at C-9 is the kinetic preference, sulphonation with sulphuric acid at 20 °C occurs to give mainly 1,5- and 1,8-anthracene disulphonic acids (Fig. 7.15). Even so, these products still suffer adverse non-bonded interactions between the sulphonic acid functions and the *peri*-hydrogen atoms, and when the reaction is repeated at 130 °C the more thermodynamically stable 2-anthracene sulphonic acid is obtained, together with 2,6- and 2,7-anthracene disulphonic acids.

7.3.2 Phenanthrene

7.3.2.1 Synthesis
There are numerous syntheses of phenanthrene, mainly due to the fact that the phenanthrene strutural unit is frequently found in complex natural products.

Naphthalene undergoes the Haworth succinoylation reaction to afford 1-naphthyl-4-oxobutanoic acid, which can be reduced under Clemmensen conditions (Zn/HgCl) to afford the corresponding butanoic acid. This product is then cyclized to 1,2,3,4-tetrahydro-1-oxophenanthrene by treatment with sulphuric acid. Reduction, dehydration, and dehydrogenation of the tetrahydrophenanthrene affords phenanthrene, whereas reactions with alkyl Grignard reagents, prior to dehydration and oxidation, gives 1-alkylphenanthrenes (Fig. 7.16).

Trans-stilbene when irradiated with ultraviolet light gives a *trans*-dihydrophenanthrene which readily aromatizes to phenanthrene in the presence of an oxidizing agent such as iodine or oxygen (eqn 7.17). The reaction can be used to cyclize derivatives of stilbine, although its application is limited to compounds which bear electronically similar groups in the two rings.

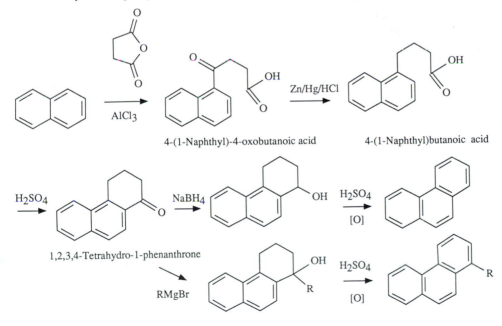

4-(1-Naphthyl)-4-oxobutanoic acid   4-(1-Naphthyl)butanoic acid

1,2,3,4-Tetrahydro-1-phenanthrone

Fig. 7.16 Synthesis of phenanthrene and 1-alkylphenanthrenes

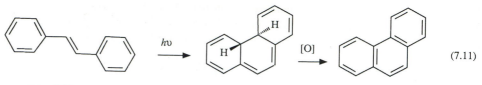

trans-Stilbene

Under photochemical conditions *trans*-stilbene is converted into *cis*-stilbene; it is this species, as its excited singlet, which cyclizes to the dihydrophenanthrene. The *trans* stereochemistry in the product is determined by the principle of the *conservation of orbital symmetry*, see Fleming (1976).

7.3.2.2 Structure and reactions

The resonance description of phenanthrene shows that the bond between C-9 and C-10 has a high degree of double bond character (Fig. 7.17). In line with this prediction, the chemistry of phenanthrene shows many similarities to that of anthracene. Addition reactions across the 9,10-bond are common, but rather less easily effected. This can be accounted for since initial attack of an electrophile (Y^+) produces a *singly* benzylic carbocation, which can either be deprotonated or quenched with available nucleophiles (X^-) (Fig. 7.18).

For anthracene the corresponding cation is *doubly* benzylic (see p. 81).

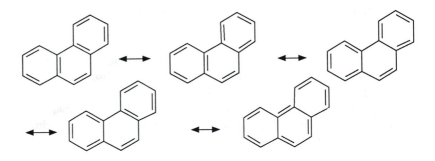

Fig. 7.17 Resonance description of phenanthrene

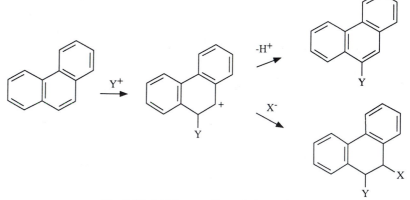

Fig. 7.18 Addition reactions of phenanthrene

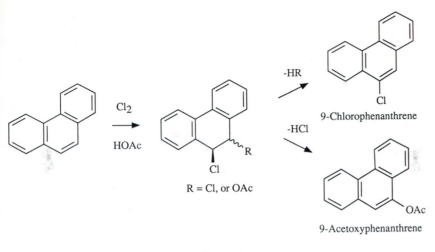

Fig. 7.19

Chlorine in acetic acid solution gives *cis*- and *trans*-9,10-dichloro-9,10-di-hydrophenanthrenes, plus *cis*- and *trans*-9-acetoxy-10-chloro-9,10-dihydrophenan-threnes. Both the dihalides and the acetoxychloro derivatives eliminate hydrogen chloride to form 9-chlorophenanthrene or 9-acetoxyphenanthrene respectively. The acetoxychloro compounds may also lose acetic acid and give 9-chlorophenanthrene (Fig. 7.19).

In the less polar solvent, carbon tetrachloride, bromine yields 9-bromophenan-threne. This reaction has been shown to be a typical aromatic substitution reaction, not one involving the above type of addition–elimination process.

Oxidation of phenanthrene with chromic acid gives 9,10-phenanthraquinone, which on reduction with lithium aluminium hydride affords *trans*-9,10-dihydroxy-9,10-dihydrophenanthrene plus 9,10-dihydroxyphenanthrene (eqn 7.12).

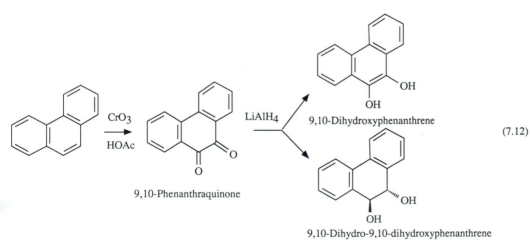

(7.12)

8 Wider aspects of aromaticity

8.1 Aromatic carbocycles and heterocycles

The Hückel ($4n+2$ π electron) rule predicts that certain planar cyclic molecules, built up through the union of sp^2 hybridized carbon atoms, should show aromatic character. In principle, however, other atoms can replace carbon provided that they contribute an appropriate number of p electrons to the π system. Pyridine, for example, is isoelectronic with benzene. The nitrogen atom is sp^2-hybridized, and of the three hybrid orbitals two are involved in bonding to the flanking carbon atoms, while the third contains a non-bonded pair of electrons. The remaining valency electron of the nitrogen atom is not hybridized, it resides in a p orbital which overlaps the p lobes of the carbon atoms, thereby forming a delocalized six π-electron system, similar to that of benzene.

Six atoms in a ring are not always required, and the five membered heterocycles pyrrole, furan, and thiophene are also aromatic molecules. Each of the four carbon atoms in these compounds contributes one p electron, but the hetero atom provides the extra two p electrons needed to make up the aromatic sextet. It follows that both pyridine and pyrrole share aromatic character. In addition pyridine is a base since it has a lone pair of electrons at the nitrogen atom which is available to capture a proton. This feature is not present in the five membered heterocycle pyrrole, which is a neutral molecule. Obviously, more than one heteroatom can be incorporated into the ring and these atoms do not necessarily need to be identical. As a result there are many aromatic heterocycles, both monocyclic and polycyclic. The subject area is immense and outside the scope of this text.

Pyridine

Pyrrole

Furan

Thiophene

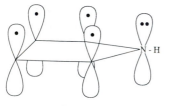

Pyridine

Pyrrole

8.2 The annulenes

Returning to carbocycles, the simplest aromatic compounds are cyclopropenium salts. These obey Hückel's rule, and exemplify the case where n = 0, so that $(4n+2) = 2$ (see p. 4). Despite the ring strain that the cyclopropenium tricyclic inherits from having formal bond angles of only 60°, cyclopropenium salts do exist and even survive in aqueous media. They represent some of the most stable carbocations known. Cyclopropenone and some of its derivatives are also stable molecules, and here the aromaticity of the ring can be interpreted if the dipolar resonance description of the carbonyl group is considered (Fig. 8.1). This places a positive charge upon the carbon atom and allows the electrons of the double bond to generate a completely delocalized two π electron system.

Taking Hückel's rule as the criterion for aromaticity, it follows that larger cyclic planar molecules should also be aromatic, provided they have the appropriate number of p electrons. Benzene is a member of a series of monocyclic compounds called annulenes, each differing from its neighbour by the presence, or absence, of an additional CH=CH unit. Each member of the series is described by a number in

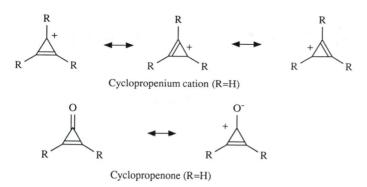

Cyclopropenium cation (R=H)

Cyclopropenone (R=H)

Fig. 8.1 Resonance descriptions of the cyclopropenium cation and cyclopropenone

square brackets which denotes the number of *p* electrons present, thus benzene is [6]annulene and cyclooctatetraene is [8]annulene. There has been much effort to synthesize larger and larger annulenes and to examine them for aromatic character, most conveniently by ^{1}H NMR spectroscopy, which determines whether or not they sustain a ring current (see p. 2).

Hückel aromaticity is not the only factor, however, and there is a competition between stability, brought about by electron delocalization, and steric effects. There are, for example, three possible isomers of [10]annulene: the all-*cis,* the mono-*trans,* and the *cis-trans-cis-cis-trans.* The first two, but not the last, have been synthesized and obtained as crystalline solids at -80 °C. However, their ^{1}H NMR spectra show only resonances in the olefinic region and it is clear that these molecules cannot achieve planarity, which is a hallmark of an aromatic compound and necessary for the effective overlap of individual p orbitals (see p. 2). Were [10]annulene to be a regular decagon, the internal C–C bond angles would be 144°, which in this case would induce more strain than could be offset by the resonance effect. Consequently the all-*cis* isomer is buckled, whereas the mono-*trans* resembles the tub structure of cyclooctatetraene.

Although ring strain is eliminated in the case of the *cis-trans-cis-cis-trans* isomer, the hydrogen atoms at the centre of the molecule point inwards and interfere sterically with one another. This non-bonded interaction is the reason for the instability of the compound, and it can be removed by bridging in the two central carbon atoms. 1,6-Methano[10]annulene, for example, can be prepared: it is stable and undergoes substitution, rather than addition, reactions (Fig. 8.2).

[14]Annulene is known, but cannot achieve planarity for similar reasons to those described for *cis-trans-cis-cis-trans*[10]annulene. In contrast, [18]annulene is suf-

Angle strain is not insurmountable and several other molecules have been made which are planar and do have large internal angles.

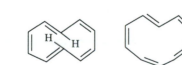

all *cis*   *cis-trans-cis-cis-trans*   mono-*trans*   1,6-Methano[10]annulene

Fig. 8.2 [10]Annulenes

[14]Annulene

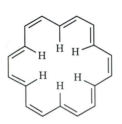

[18]Annulene

ficiently large to allow the internal hydrogen atoms enough space to be accommodated without severe interactions; as a result, the molecule is planar.

The ^{1}H NMR spectrum of [18]annulene shows two resonances, one due to the signals of the protons on the outside of the ring, at a chemical shift value of 9.0 ppm, and the other due to the resonances of the inner proton signals, at the very high field (shielded) position of -3.0 ppm (i.e. above the internal standard tetramethylsilane at 0 ppm). This evidence confirms the aromaticity of [18]annulene and indicates that a ring current is established in this molecule. The outer protons lie in the deshielding zone created by this ring current, whereas the inner proton are orientated directly into the shielding region of the induced field.

References

Fleming, I. (1976). *Frontier orbitals and organic reactions*. Wiley, London.
Hammond, G. S. (1955). *J. Am. Chem. Soc.* **77**, 334.
Hine, J. (1977). *Adv. Phys. Org. Chem.* **15**, 1.
March, J. (1985). *Advanced organic chemistry*, 3rd edn, p. 701. McGraw Hill.

Exercises

1. Explain the presence of *o*-, *m*- and *p*- nitroanilines in the reaction mixture formed when aniline is reacted with nitric and sulphuric acids.

2. Predict the nitration products of (a) toluene, (b) benzonitrile (phenyl cyanide), (c) bromobenzene, (d) acetanilide (*N*-acetylaniline), (e) trifluorobenzene.

3. Draw a reaction profile for the sulphonation of benzene, which illustrates the fact that the reaction can be reversed at elevated temperature and shows a small hydrogen-deuterium isotope effect ($K_H / K_D \approx 2$).

4. Rank the following compounds in increasing rate of monobromination and suggest the major isomer in each case. (a) 4-nitrophenol, (b) 3-methylphenol, (c) 3-nitrophenol, and (d) 4-methylphenol.

5. Why do phenols require moderately basic conditions for coupling reactions with diazonium salts?

6. Account for the fact that aryldiazonium salts are more stable than their alkyl counterparts.

7. Rank in term of reactivity towards nucleophiles 2,4,6-trinitro-, 2,4-dinitro, 3-nitro, 4-nitro-chlorobenzenes.

8. Account for the fact that 1,3,5-trimethylbenzene when reacted with ethyl fluoride/boron trifluoride at -80 °C yields a salt which when it is warmed to -15 °C affords hydrogen fluoride, boron trifluoride and 2,4,6-trimethylethylbenzene.

9. Draw out the structure of nitrosobenzene, including electron pairs, and show why it directs *ortho* and *para* in electrophilic reactions. Will the rate of substitution be faster or slower than that of aniline?

10. Place in order of acidity phenol, benzoic acid, and cyclohexanol. Which is most acidic 3-nitrophenol or 4-nitrophenol?

11. Suggest why salicylic acid (2-hydroxybenzoic acid) is more acidic than its *meta* and *para* isomers, but less so than 2,6-dihydroxybenzoic acid.

12. Show why 4-methylchlorobenzene when reacted with sodium amide in liquid ammonia gives both 3- and 4-methylanilines.

13. Account for the fact that when 2-aminobenzoic acid is diazotized and the product is heated in methanol carbon dioxide and nitrogen are lost and methoxybenzene (anisole) is formed.

14. Discuss the fact that toluene when reacted with chlorine in sunlight affords benzyl chloride, whereas in the presence of iron (III) chloride *o*- and *p*-chlorotoluenes are formed.

15.   Draw arrow mechanisms to show why 2-(4-methoxyphenyl)-2-chloropropane is hydrolysed to the corresponding alcohol much faster than 2-phenyl-2-chloropropane.

16.   Comment upon the fact that toluene on electrophilic substitution affords *o*- and *p*-substituted toluenes whereas *meta*- products arise from trichloromethylbenzene (benzotrichloride).

17.   Predict the product(s) from the reaction of benzenediazonium chloride with (a) aniline, (b) *N*,*N*-dimethylaniline, (c) *N*,*N*-2,6-tetramethylaniline.

18.   Place in order of basicity aniline, acetanilide (*N*-acetylaniline), and cyclohexylamine.

19.   Why should 2,4,6-trinitroaniline be 40 000 times less basic than *N*,*N*-dimethyl-2,4,6-trinitroaniline, but aniline and *N*,*N*-dimethylaniline are almost equally basic?

20.   Draw mechanisms for the reaction of benzaldehyde with (a) HCN/KCN, (b) propanone/NaOEt, (c) HCHO/NaOH.

21.   Why should 1-bromocyclohepta-2,4,6-triene (tropylium bromide) be freely soluble in water, whereas benzyl bromide is not?

22.   Explain why naphthalene is sulphonated at low temperature at C-1, whereas at higher temperatures 2-naphthalene sulphonic acid is produced.

23.   9,10-Diphenylanthracene reacts with oxygen in ultraviolet light to give an adduct $C_{26}H_{18}O_2$. What is its structure and why does it form?

24.   What is the structure of the compound formed by the reaction of phenanthrene and benzyne?

25.   Why is the reaction product the same when benzene is reacted with either chloropropane or propene in the presence of aluminium (III) chloride?

Index